T0305716

Advanced Modelling and Simulation in the Chemical and Biochemical Process Industry

Advanced Modelling and Simulation in the Chemical and Biochemical Process Industry explores modelling and simulation of chemical and biochemical processes at the industrial scale using a variety of approaches. Particular attention is devoted to simulations in different scales, which help achieve a wide-spectrum and more efficient analysis of several problems, ranging from the design of novel materials to the optimization of industrial processes as a function of the operating conditions. This book not only covers optimization with experimental data but also offers readers a thorough understanding and analysis of different parameters of a whole process stream.

- Covers a wide range of advanced modelling and simulation of chemical technologies: ab initio, atomistic molecular dynamics (MD), Lattice-Boltzmann (LB), dissipative particle dynamics (DPD), computational fluid dynamics (CFD), and finite element (FEM)
- Addresses issues associated with process control in different phases of the chemical industry
- Features modelling approaches that allow the design of novel processes/ materials in a faster and more reliable way

This book will be of interest to researchers and advanced readers in chemical, biochemical, environmental, and materials engineering and industrial chemistry.

Sudip Chakraborty earned a PhD in chemical engineering at the University of Calabria, Italy. He is a faculty member in the Department of Informatics, Modeling, Electronics, and Systems Engineering, University of Calabria, Italy. His major fields of interest include membrane separation, plasmonic nanoparticles, composite materials, energy, and process intensification. He has an h-index of 42 and has published more than 130 research publications in international journals, books, and conference proceedings.

Stefano Curcio is the Director of the Department of Computer Engineering, Modeling, Electronics and Systems and a full Professor at the University of Calabria. He is a visiting researcher at the European Organization for Nuclear Research (CERN), Geneva. He has written more than 75 papers that have been published in international peer-reviewed scientific journals, ten book chapters, one patent, and more than 120 presentations in international conferences and published in conference proceedings.

Advanced Modelling and Simulation in the Chemical and Biochemical Process Industry

Edited by
Sudip Chakraborty and Stefano Curcio

CRC Press
Taylor & Francis Group
Boca Raton London New York

CRC Press is an imprint of the
Taylor & Francis Group, an **informa** business

Designed cover image: © 2025 Shutterstock, PopTika

First edition published 2025
by CRC Press
2385 NW Executive Center Drive, Suite 320, Boca Raton FL 33431

and by CRC Press
4 Park Square, Milton Park, Abingdon, Oxon, OX14 4RN

CRC Press is an imprint of Taylor & Francis Group, LLC

© 2025 selection and editorial matter, Sudip Chakraborty and Stefano Curcio; individual chapters, the contributors

ISBN: 978-1-032-56369-5 (hbk)
ISBN: 978-1-032-56370-1 (pbk)
ISBN: 978-1-003-43518-1 (ebk)

DOI: 10.1201/9781003435181

Typeset in Times
by KnowledgeWorks Global Ltd.

Contents

Preface

This is an introductory textbook on simulation-based process optimization that is intended for undergraduates and graduate students in engineering. The principal coverage includes but is not limited to the use of simulation and optimization processes in industrial applications. Although a lot of operations research and optimization books touch on modelling or optimization techniques, the shortfall is that they usually don't get in coverage, which is reflected in their minimal application to problems in the real world. Illustrating the important influence of modelling on the decision-making process, *Advanced Modelling and Simulation in the Chemical and Biochemical Process Industry* helps you come to grips with a wide range of modelling techniques in chemical and biochemical process industries. The book adopts many examples of design scenarios as context for curating sample problems. This will help students relate desktop problem-solving to tackling real-world problems. Succinct yet rigorous, with over 100 pages of problems and corresponding worked solutions presented in detail, the book is ideal for students of engineering, applied science, and market analysis.

Highlighting the modelling aspects of optimization problems in various chemical and biochemical process, the authors present the techniques in a clear and straightforward manner, illustrated by a few case studies. They provide and analyse the formulation and modelling of several well-known theoretical and practical problems and touch on solution approaches. The book demonstrates the use of optimization packages through the solution of various mathematical models and provides an interpretation of some of those solutions. It presents the practical aspects and difficulties of problem-solving and solution implementation and studies. The book also discusses the use of available software packages in solving optimization models without going into difficult mathematical details and complex solution methodologies. A wide range of methods, classic theoretical, and practical problems is considered with data collection and input, preparation, and practical issues of modelling, model solving, and validation. The authors draw directly from their experience to provide lessons learned when applying modelling techniques to practical problem-solving in chemical and biochemical process industries.

This book covers using simulation and optimization, a key area of operations research, which has been applied to virtually every chemical and biochemical process industry. With the rapid rise of interest in simulation and data analytics, this book is timely. Working technology and business professionals need an awareness of simulation and optimization tools. Philosophically, the book emphasizes the creation of formulations before going into implementation in industry. This book:

- Provides and explains case studies so examples are relatively clear and self-contained
- Underlines the need to create formulations before implementing
- Focuses on application rather than algorithmic details
- Embodies the philosophy of reproducible research
- Emphasizes specific optimization modelling based on practical application

With the information in each chapter, readers can adapt their own applications. This book can be used for graduate and undergraduate courses for students without a background in optimization and with varying mathematical or simulation backgrounds.

And finally, our family for bearing with us and supporting us all these years through thick and thin in professional and personal life. Thank you all from the bottom of my heart. Although we have been careful in developing the manuscript, we wouldn't be surprised if you find errors in the book. We are solely responsible for those; please do not hesitate to bring them to our attention in case if you find any.

Contributors

Hernan D. Alvarez
Universidad Nacional de Colombia
Medellín, Colombia

Omotola Babajide
Cape Peninsula University of
  Technology
Cape Town, South Africa

Vena Pearl Bongolan
University of the Philippines
Quezon City, Philippines

Otmane Boudouch
Université Chouaib Doukkali
El Jadida, Morocco

Roy Vincent L. Canseco
University of the Philippines
Quezon City, Philippines

Alfredo Cassano
Istituto per la Tecnologia delle
  Membrane
Consiglio Nazionale delle Ricerche
Rende, Italy

Sudip Chakraborty
Università della Calabria
Rende, Italy

Carmela Conidi
Istituto per la Tecnologia delle
  Membrane
Consiglio Nazionale delle Ricerche
Rende, Italy

Stefano Curcio
Università della Calabria
Rende, Italy

Michael O. Daramola
University of the Witwatersrand
Johannesburg, South Africa

Sirshendu De
Indian Institute of Technology
Kharagpur, India

Rizalinda L. de Leon
University of the Philippines
Quezon City, Philippines

Giorgio De Luca
Istituto per la Tecnologia delle
  Membrane
Consiglio Nazionale delle Ricerche
Rende, Italy

Radouane El Amri
Université Sultan Moulay Slimane
Beni Mellal, Morocco

Reda Elkacmi
Université Sultan Moulay Slimane
Beni Mellal, Morocco

Madhumita Maitra
St. Xavier's College
Kolkata, India

Sourav Mondal
Indian Institute of Technology
Kharagpur, India

Arijit Nath
Hungarian University of Agriculture
  and Life Sciences
Budapest, Hungary

Francesco Petrosino
Università della Calabria
Rende, Italy

Santanu Sarkar
Tata Steel Ltd.
Jamshedpur, India

Joseph Yap IV
University of the Philippines
Quezon City, The Philippines

Rajaa Zahnoune
Université Sultan Moulay Slimane
Beni Mellal, Morocco

1 Modeling of Gel Controlling Membrane Filtration in Fruit Juice Processing

Sourav Mondal, Carmela Conidi,
Alfredo Cassano, and Sirshendu De

1.1 INTRODUCTION

Fruit juices are recognized as important components of the human diet, providing a range of key nutrients such as vitamins, antioxidants and minerals which are important for their role in preventing chronic diseases such as cancer, cardiovascular and neurological disorders (Kaur and Kapoor, 2001). Traditional methods involved in fruit juice processing (extraction, pasteurization, concentration, packaging, etc.) have a significant impact on the organoleptic and nutritional properties of the final product. At the same time, they are characterized by several drawbacks in terms of environmental protection. For instance, conventional juice clarification processes are based on the use of fining agents (gelatine, bentonite, silica sol, diatomaceous earth, etc.), which create serious problems of environmental impact due to their disposal. Similarly, the concentration of fruit juices, aiming at ensuring a longer storage life and an easier transportation, is performed by thermal evaporation methods, resulting in color degradation and reduction of most thermally sensitive compounds, with a consequent remarkable qualitative decline (Braddock and Goodrich, 2003).

Product quality improvement and energy savings have guided the development of minimal processing techniques. In this view, membrane separation processes have emerged as a valid alternative to traditional thermal techniques for fruit juice clarification and concentration. These processes involve no phase change or chemical agents and can be operated at ambient temperatures. These features are key factors in the production of additive-free fruit juices with natural fresh tastes. In addition, membrane processes are characterized by less manpower requirement, greater efficiency and shorter processing time than conventional filtration. Consequently, the operational costs of using membrane processes are significantly lower than those of conventional processes (Nunes and Peinemann, 2001).

Microfiltration (MF) and ultrafiltration (UF) are pressure-driven membrane operations commonly used in fruit juice processing (Bhattacharjee et al., 2017). These processes are able to separate particles in the approximate size ranges of 1–100 and

0.1–10 μm, respectively. In particular, UF membranes are able to retain large species such as microorganisms, lipids, proteins and colloids while small solutes such as vitamins, salts and sugars flow together with water. Commercial clarification processes based on the use of UF membranes have been implemented on an industrial scale for fruit juices such as apple and orange (Girard and Fukumoto, 2000). They are able to retain large species such as microorganisms, lipids, proteins and colloids while small solutes such as vitamins, salts and sugars flow together with water. Typical advantages over conventional fruit juice clarification are: increased juice yield; possibility of operating in a single step; possibility of avoiding the use of gelatins, adsorbents and other filtration aids; reduction in enzyme utilization, easy cleaning and maintenance of the equipment; elimination of needs for pasteurization; better juice clarity; and reduction of filtration times and waste products (DasGupta and Sarkar, 2012).

Despite the many benefits of membrane clarification, the performance of this operation is affected by a permeate flux decline with time, caused by accumulation of juice components at the membrane surface. This phenomenon is known as membrane fouling; it may occur due to a concentration polarization layer development over the membrane surface, the formation of a cake layer and/or a blockage of the membrane pores (Field et al., 1995). The pore blocking can be further characterized by complete, intermediate and standard pore blocking. Membrane fouling is a key factor affecting the economic and commercial viability of a membrane system, which essentially depends on the permeate fluxes obtained and their stability with time (Hojjatpanah et al., 2011). It controls the frequency of cleaning, the lifetime of the membrane, the area needed for separation and, indirectly, the costs, design and operating parameters of membrane plants (Mohammad et al., 2012). Therefore, the identification and quantification of the prevalent fouling mechanism and efforts to minimize its effects during a continuous filtration process are extremely important (Mondal et al., 2011a).

In most filtration processes, batch mode is often used, since the permeate is the preferred product. For efficient design of large-scale systems, prediction and detailed understanding of the mass transfer phenomena with coupled fluid flow are important. The relevant flow configuration and flow regimes play a critical role in modeling the process performance. The mass transfer coefficient is generally calculated from the Sherwood number relations, derived from heat and mass transfer analogies. However, these correlations fail to take into account the developing mass transfer boundary layer on the hydrodynamics of the flow regime.

Fruit juices are a complex mixture of several components in the solution. Since they contain high molecular weight proteins, fibers, cellulose, etc., the separation process is primarily gel layer-controlled (Barros et al., 2003). There have been several attempts reported in literature to apply resistance in series and pore blocking models to predict the flux decline behavior (Vladisavljević et al., 2003; Rai et al., 2006; Cassano et al., 2007; Rai et al., 2007). However, such models lack the physical understanding of the process necessary to capture the variation in feed characteristics and are mostly empirical in nature. The flow characteristics in hollow fiber membrane modules have significant effects on the mass transfer analysis.

An important phenomenon leading to the decline in flux is concentration polarization (Sablani et al., 2001). A simple description of concentration polarization is obtained from a stagnant film model, used by Sherwood et al. (1965) to analyze reverse osmosis. Many researchers (Opong and Zydney, 1991; Zydney, 1997; Johnston and Deen, 1999) have used the stagnant film model that considers a thin layer of solute of uniform thickness adhered to the membrane surface, leading to a one-dimensional problem in which the solute concentration depends only on distance from the membrane surface. To overcome this problem, a detailed numerical solution of the governing momentum and solute mass balance equations with pertinent boundary conditions may be used (Kleinstreuer and Paller, 1983; Bouchard et al., 1994; De and Bhattacharya, 1997a). However, these studies do not incorporate the effects of fluid rheology and involve inherent complexities and rigorous computational requirements, rendering them unattractive and not useful for fruit juice clarification applications.

Detailed studies related to two-dimensional concentration fields for laminar crossflow UF in tubes or parallel-plate channels (Shen and Probstein, 1977; Gill et al., 1988; Denisov, 1999; Madireddi et al., 1999) and spiral-wound membrane modules (Kozinski and Lightfoot, 1971) have been reported in literature. Field and Aimar have modified Leveque's relation for laminar flow in the rectangular channel by using a viscosity correction factor (Field and Aimar, 1993). However, the effects of suction were not considered in their study, which was incorporated later by De and Bhattacharya (1997a). Sherwood number relationships incorporating the effects of suction (in the presence of a membrane) for laminar flow in rectangular, radial, and tubular geometries, have been formulated starting from first principles (De and Bhattacharya, 1997a). However, these studies include the osmotic pressure-controlled filtration only and the effect of developing mass transfer boundary layer.

It has been shown that due to concentration polarization, the variation of the physical properties with concentration is significant in the performance of UF and subsequent development of the boundary layer, especially in a gel layer controlling case (Gill et al., 1988; De and Bhattacharya, 1999; Bowen and Williams, 2001). During filtration of high molecular weight proteins, polymers, paint, clay, etc., a highly viscous layer is formed over the membrane surface (commonly known as gel concentration), and it obeys the classical gel filtration theory. The primitive gel layer model is derived from conventional film theory (Blatt et al., 1970), considering a uniform mass transfer boundary layer thickness instead of a developing boundary layer, which is fundamentally correct. Moreover, the viscosity of the solution is a strong function of the solute concentration, and it varies significantly within the mass transfer boundary layer. The concentration of the gel layer is much higher than the bulk concentration because of the concentration polarization, affecting the viscosity remarkably. This variation of viscosity as a function of concentration was not included in the film model. Probstein et al. (1978) proposed a two-dimensional model, developing a mass transfer boundary layer under laminar flow condition in a rectangular channel, overcoming the limitation of the constant thickness boundary layer. Clarification of fruit juices by UF has been found to be gel controlling in many occasions due to presence of protein, cell debris, cellulose, etc. (Sarkar et al., 2008; Rai et al., 2010; Mondal et al., 2012).

The present chapter attempts to quantify the flux decline behavior during crossflow UF of fruit juices through hollow fiber membranes. The mathematical analysis was based on the first principles using the integral method of solution to compute the developing concentration boundary layer over the gel layer. The analysis was extended for the batch mode of filtration, and the profiles of permeate flux were computed for different operating conditions in the predictive mode. The simulated results were also compared with the experimental data of fruit juices.

Some of the key and distinguishable features of the present modeling approach are summarized here:

1. The present model captures the essential underlying physics of the species transport in the concentration boundary layer and gel formation, which is not described well by many semi-empirical fouling models.
2. Most fouling models are based on Hermia's (1982) and Field et al.'s (1995) models based on the constant pressure filtration, which does not take into account of the increase in feed concentration during the batch mode of filtration and the effects of the channel narrowing due to gel formation.
3. The values of the unknown physical properties and process constants are realistic and are not fitting parameters which are obtained by regression in other black box models.
4. The results of the model can be directly interpreted on the effect of the process parameters (transmembrane pressure drop, flow rate, etc.) and thus are useful in understanding the interplay of the operating conditions and controlling such systems in practice.
5. Finally, many existing fouling models are not suitable in incorporating the fluid rheological effects on the overall mass transport phenomena.

Gel layer thickness is difficult to measure experimentally during filtration, and few studies on the in situ measurements of the gel layer during unstirred batch membrane filtration are reported in literature (Chen et al., 2004; Guell et al., 2009). To the best of the authors' knowledge, there is no analytical or instrumentation facility available to date to directly measure the thickness of the gel layer inside the hollow fiber during filtration. The classical gel layer theory used to predict the gel thickness was based on the solute mass balance at the fluid-gel interface and requires the knowledge of the gel layer concentration and permeate flux (Blatt et al., 1970; Porter, 1972; Belfort et al., 1994). Thus, it is impossible to predict the permeate flux or the gel concentration from this theory. Also, this is based on the assumption of the constant concentration of the boundary layer developed over the membrane surface. There has been an attempt to include the shear-induced diffusion of the particulate suspension (Davis, 1992) and solve the transient 1D species transport equation (Karode, 2001) as an improvement to the existing models. One of the prevalent theories in this regard is the application of the Happel cell model (Happel and Brenner, 1965) based on particle flux conservation (Song and Elimelech, 1995). The gel is considered to be a concentrated particulate suspension of non-interacting uniform hard spheres. However, this is a 1D model and does not include the effects of the forced convection due to crossflow. Also, the time-dependent growth of the gel layer is not accounted for. The model presented in this chapter is fundamentally based on the 2D concentration boundary layer analysis. This highlights the applicability and strength of the model, as it can predict the profile of gel layer thickness from the knowledge of the operating conditions and physical parameters, which was not possible otherwise.

1.2 CHALLENGES IN FRUIT JUICE FILTRATION MODELING

Clarification of fruit juices using membrane technology is an advanced treatment process that has multiple advantages over the traditional methods. With the growth of the market potential in processed fruit juices, membrane filtration tends to have the leading market share. Thus, scaling up to large processing units is quite important, and therefore mathematical modeling is quite essential and necessary. However, there are couple of challenges associated with the transport phenomena of the process:

i. Rheology of the fruit juice is critical:
   The viscosity of the fruit juice used in the membrane system is an important parameter. Viscosity of a solution is affected considerably by the changes in the solute concentration. On account of the formation of the concentration boundary layer in the system, the viscosity changes in the flow channel above the membrane surface are substantial. Therefore, including the viscosity variation effect in the mass transfer analysis is significant.

   Further, viscosity of the fruit juice solution is dependent on the shear rate (non-Newtonian effects), as well as the temperature of the system. So inherently such factors also play a role in the filtration performance manifested through the changes in the operating conditions.
ii. Gel layer controlled separation:
   Since the fruit juices contain several high-molecular-weight bioproducts (i.e. cellulose, hemi-cellulose, fibers, carbohydrates, pectins and cell-debris) they are typically handled using MF or UF processes. As the molecular weights of these solutes are high, they tend to produce a gel-forming material, which ultimately leads to the formation of the gel layer over the membrane, acting as a dynamic membrane. The resistance of the gel layer is generally the dominant one and controls the overall filtration rate. It must be noted that the mathematical model should be equipped to focus on the problem related to the dynamics of the gel, which is purely a mass transfer effect.
iii. Unknown physical properties of the system:
   Fruit juices are composed of different biomolecules and bioproducts, which have individual physical properties and different degrees of separation potential. For a mechanistic model of a physical process, knowledge of the different physical properties is quite crucial. However, it is often difficult to determine the different physico-chemical properties of the fruit juice, particularly with respect to the changing composition due to the separation. Examples of such properties include diffusivity, gel concentration, etc. Therefore, this is an added complexity in developing such models and often results relied on the optimization of the output profiles of the controlled experiments based on the defined operating conditions.

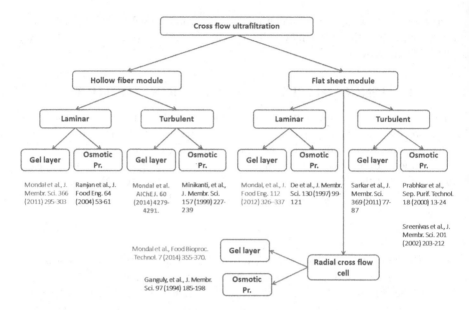

FIGURE 1.1 Flowsheet showing the different possible classifications in the mass transport-based membrane filtration models under crossflow operation mode.

1.3 CLASSIFICATION OF MASS TRANSFER-BASED MEMBRANE MODELS

Membrane filtration in practice is typically performed in crossflow mode. The different mass transport-based models can be classified based on the crossflow configuration, flow regimes and mechanism of solute transport. A summary of different models along with the associated references is illustrated in Figure 1.1.

1.4 FRUIT JUICE FILTRATION USING HOLLOW FIBER MEMBRANES

1.4.1 FRUIT JUICE FILTRATION MODEL IN THE LAMINAR REGIME

Hollow fibers are routinely used in fruit juice clarification processes because of their compact size, high filtration area in small floor space and easy scaling of the operation. A detailed mass transport-based modeling of the concentration boundary layer formed over the membrane surface during the UF of kiwi fruit juice has been reported by Mondal et al. (2011b). The following assumptions were considered in the model:

1. The diffusivity and viscosity do not depend on the concentration of the suspended particles.
2. There are no particle-wall and particle-particle interactions.
3. The gel concentration is below the solubility limit, so the continuum equations are still valid.

4. The density of the feed and permeate are the same, and the fluid is incompressible in nature.
5. The fluid flow is in the laminar zone and fully developed.

The information on the mass transfer coefficient and the permeate fluxes can be obtained from the knowledge of the concentration profiles. The model approach in both the total recycle and batch concentration mode of operation is discussed.

In the total recycle mode of operation, both the permeate and retentate streams are recycled back to the feed tank so that a steady state is attained with fixed concentration of the feed. This is a popular mode of operation in any membrane-based process in order to evaluate the effects of operating parameters (feed concentration, crossflow velocity, transmembrane pressure drop) on steady-state permeate flux and permeate quality. During the UF of fruit juice, high molecular weight solutes are transported toward the membrane wall, forming a gel layer over the membrane surface. The gel concentration is assumed to be constant within this layer. Therefore, there exists a concentration boundary layer next to the gel layer with a variation in concentration from bulk to gel concentration. The convective-diffusive flux equation for the gel-forming solute within a concentration boundary layer is given for a tubular module as,

$$u\frac{\partial C}{\partial x} + v\frac{\partial C}{\partial r} = \frac{\partial}{\partial r}\left(D\frac{\partial C}{\partial r}\right) \tag{1.1}$$

Since the thickness of the concentration boundary layer is much less than the radius of the tube, the curvature effect is entirely lost and Eq. (1.1) is reduced to a planar coordinate by defining $y = R - r$, where y is the distance from the membrane surface (tube wall) (Lévêque, 1928). For easy understanding and reference, a schematic of the geometry is shown in Figure 1.2.

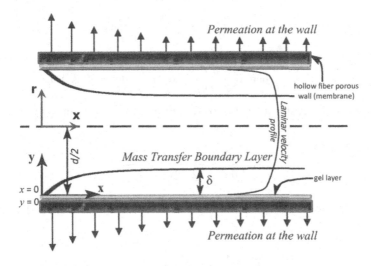

FIGURE 1.2   Schematic of the geometrical setup of the problem.

Under this coordinate system, the above equation becomes

$$u\frac{\partial C}{\partial x}+v\frac{\partial C}{\partial y}=D\frac{\partial^2 C}{\partial y^2} \tag{1.2}$$

It may be noted that the crossflow velocity in a membrane module is generally 5–6 orders of magnitude greater than the permeation velocity. Thus, it can be assumed that the parabolic velocity profile (under laminar flow regime) inside the tubular module remains undisturbed due to permeation in the wall. The velocity in the tube is given by the Poiseuille law,

$$u = 2u_0\left[1-\left(\frac{r}{R}\right)^2\right] \tag{1.3}$$

Since $\frac{y}{R}\ll 1$, the x-component velocity profile simplifies to $u \sim 4u_0\frac{y}{R}$. Since, the concentration boundary layer thickness is extremely small, it can be assumed that the y-component velocity is equal to the permeation velocity at the wall (De et al., 1997). Therefore, the y-component velocity becomes $v = -v_w$. Inserting the velocity profiles into the Eq. (1.2), the following equation is obtained,

$$4u_0\frac{y}{R}\frac{\partial C}{\partial x}-v_w\frac{\partial C}{\partial y}=D\frac{\partial^2 C}{\partial y^2} \tag{1.4}$$

Using $x^* = x/L$; $y^* = y/R$; $C^* = C/C_0$, the above equation is non-dimensionalized as,

$$\frac{u_0 d^2}{DL}y^*\frac{\partial C^*}{\partial x^*}-\frac{v_w d}{2D}\frac{\partial C^*}{\partial y^*}=\frac{\partial^2 C^*}{\partial y^{*2}} \tag{1.5}$$

where, $d = 2R$. It is noted here that $\frac{u_0 d^2}{DL}=Re.Sc.\frac{d}{L}$ (denoted as A, henceforth) and $\frac{v_w d}{D}$ is the non-dimensional flux (denoted as Pe_w). So, Eq. (1.5) can be rearranged as,

$$Ay^*\frac{\partial C^*}{\partial x^*}-\frac{Pe_w}{2}\frac{\partial C^*}{\partial y^*}=\frac{\partial^2 C^*}{\partial y^{*2}} \tag{1.6}$$

A quadratic concentration profile within the concentration boundary profile is assumed as

$$C^* = \frac{C}{C_0}=a_1+a_2 y^*+a_3 y^{*2} \tag{1.7}$$

where a_1, a_2 and a_3 are the constant coefficients. Using the following boundary conditions, the constants of Eq. (1.7) are evaluated,

$$\text{at} \quad y^* = 0, \ C^* = C_g^* = C_g/C_0 \tag{1.8}$$

$$\text{at} \quad y^* = \delta^*, \ C = C_0; \ C^* = 1 \tag{1.9}$$

$$\text{at} \quad y^* = \delta^*, \ \frac{\partial C^*}{\partial y^*} = 0 \tag{1.10}$$

where δ^* is the non-dimensional thickness of the concentration boundary layer and C_g is the gel layer concentration. Using the boundary conditions Eqs. (1.8–1.10) the constants a_1, a_2 and a_3 are evaluated and the Eq. (1.7) is rewritten as

$$C^* = C_g^* - 2\left(C_g^* - 1\right)\left(\frac{y^*}{\delta^*}\right) + \left(C_g^* - 1\right)\left(\frac{y^*}{\delta^*}\right)^2 \tag{1.11}$$

Now $\dfrac{\partial C^*}{\partial x^*}$, $\dfrac{\partial C^*}{\partial y^*}$ and $\dfrac{\partial^2 C^*}{\partial y^{*2}}$ in Eq. (1.6) are evaluated using Eq. (1.11). These partial derivatives are established in Eq. (1.6), and after simplification the following parameter is obtained,

$$A\left(\frac{y^{*2}}{\delta^{*2}} - \frac{y^{*3}}{\delta^{*3}}\right)\frac{d\delta^*}{dx^*} - \frac{Pe_w}{2}\left(\frac{y^*}{\delta^{*2}} - \frac{1}{\delta^*}\right) = \frac{1}{\delta^{*2}} \tag{1.12}$$

Taking the zeroth moment of above equation by multiplying both sides by dy^* and integrating across the boundary layer thickness from 0 to δ^*,

$$A\left(\frac{d\delta^*}{dx^*}\right)\int_0^{\delta^*}\left(\frac{y^{*2}}{\delta^{*2}} - \frac{y^{*3}}{\delta^{*3}}\right)dy^* - \frac{Pe_w}{2}\int_0^{\delta^*}\left(\frac{y^*}{\delta^{*2}} - \frac{1}{\delta^*}\right)dy^* = \frac{1}{\delta^{*2}}\int_0^{\delta^*}dy^* \tag{1.13}$$

On solving the above integral, the following equation is obtained:

$$\frac{A\delta^{*2}}{12}\left(\frac{d\delta^*}{dx^*}\right) + \frac{Pe_w\delta^*}{4} = 1 \tag{1.14}$$

Now considering a steady-state mass balance over the membrane surface ($y = 0$), the following equation is obtained,

$$v_w C_g + D\frac{dC}{dy} = 0 \tag{1.15}$$

The non-dimensional version of the above equation is,

$$\frac{1}{2}Pe_w C_g^* + \frac{dC^*}{dy^*} = 0 \tag{1.16}$$

$\dfrac{\partial C^*}{\partial y^*}$ is evaluated from Eq. (1.11) and is substituted in the above equation to obtain

$$Pe_w \delta^* = 4\left(\frac{C_g^* - 1}{C_g^*}\right) \tag{1.17}$$

Replacing this value of $Pe_w \delta^*$ in Eq. (1.14) results in the following governing equation of concentration boundary layer thickness,

$$A\frac{\delta^{*2}}{12}\frac{d\delta^*}{dx^*} = \frac{1}{C_g^*} \tag{1.18}$$

Integration of the above equation leads to the profile of concentration boundary layer thickness with x^* as,

$$\delta^* = \left(\frac{36}{AC_g^*}\right)^{1/3} x^{*1/3} \tag{1.19}$$

1.4.1.1   Estimation of the Mass Transfer Coefficient

The definition of mass transfer coefficient (k) can be written as,

$$k\left(C_g - C_0\right) = -D\frac{\partial C}{\partial y}\bigg|_{y=0} \tag{1.20}$$

Non–dimensionalizing the above equation and substituting $\dfrac{\partial C^*}{\partial y^*}\bigg|_{y^*=0}$ from Eq. (1.7) leads to the following expression of the Sherwood number,

$$Sh = \frac{4}{\delta^*} \tag{1.21}$$

Substituting the profile of δ^* from Eq. (1.19), the expression of the Sherwood number becomes,

$$Sh\left(x^*\right) = 4\left(\frac{AC_g^*}{36}\right)^{1/3} x^{-1/3} \tag{1.22}$$

The length-averaged Sherwood number thus becomes

$$\overline{Sh_L} = \int_0^1 Sh(x^*)dx^* = 1.816\left(Re.Sc.\frac{d}{L} \right)^{1/3} C_g^{*1/3} \tag{1.23}$$

It is to be noted that the co-efficient in the Sherwood number relation using the classical Leveque solution is 1.86 for rectangular and 1.62 for tubular geometry (van den Berg et al., 1989). It may also be noted that these coefficients remain unaltered in case of incorporation of the Sieder Tate correction factor (Incropera and DeWitt, 1996). However, the coefficient increases if one includes the developing mass-transfer boundary layer. In case of rectangular geometry, the coefficient increases from 1.85 to 2.10 (Probstein et al., 1978). In the case of hollow fibers, the Sherwood number coefficient increases from 1.62 to 1.816 as shown here.

Since the gel layer concentration is several orders of magnitude higher than the bulk concentration, it is obvious that the viscosity variation within concentration boundary layers is significant. Therefore, a Sieder-Tate viscosity correction factor is included in the expression of the Sherwood number (Eq. 1.23) as (van den Berg et al., 1989),

$$\overline{Sh_L} = 1.816\left(Re.Sc.\frac{d}{L} \right)^{1/3} C_g^{*1/3}\left(\frac{\mu_m}{\mu_w} \right)^{0.14} \tag{1.24}$$

where, μ_m is the viscosity at the mean concentration of the solution within the concentration boundary layer and μ_w is the viscosity at the wall concentration (C_g).

Assuming the exponential variation of viscosity as,

$$\mu = \mu_0\, e^{\alpha C} \tag{1.25}$$

the viscosity at the wall becomes,

$$\mu_w = \mu_0\, e^{\alpha C_g} \tag{1.26}$$

In evaluating μ_m, we have to first calculate the mean concentration C_m in the concentration boundary layer. The expression of C_m^* is $C_m^* = \frac{1}{\delta^*}\int_0^{\delta^*} C^*\, dy^*$. Substituting concentration profile for Eq. (1.11), the following value of C_m^* is obtained,

$$C_m^* = \frac{1}{3}\left(C_g^* + 2 \right) \tag{1.27}$$

Hence using Eqs. (1.26) and (1.27), the following result is obtained,

$$\frac{\mu_m}{\mu_w} = e^{-\frac{2}{3}\alpha C_0\left(C_g^* - 1 \right)} \tag{1.28}$$

Substituting the value of viscosity correction factor $\dfrac{\mu_m}{\mu_w}$ in Eq. (1.24), the Sherwood number relation is modified to,

$$\overline{Sh_L} = 1.816 \left(Re.Sc\frac{d}{L} \right)^{\frac{1}{3}} \left(e^{-\frac{2}{3}\alpha C_0 \left(C_g^* - 1 \right)} \right)^{0.14} C_g^{*1/3} \tag{1.29}$$

1.4.1.2 Introduction of Temperature Effect on the Sherwood Number

According to the Stokes-Einstein equation, the solute diffusivity is directly proportional to temperature and inversely proportional to viscosity as $D \, \alpha \, \dfrac{T}{\mu}$ (Cussler, 1998). As viscosity varies with temperature inversely, so $\mu \, \alpha \, \dfrac{1}{T^m}$. Hence, the variation of diffusivity with temperature is quantified as $\dfrac{D}{D_0} = \left(\dfrac{T}{T_0} \right)^{m+1}$. Therefore, using Eq. (1.29), the following equation for the average Sherwood number is obtained.

$$Sh_L = 1.816 \left(Re.Sc\frac{d}{L} \right)^{\frac{1}{3}} \left(e^{-\frac{2}{3}\alpha C_0 \left(C_g^* - 1 \right)} \right)^{0.14} \left(\frac{T_0}{T} \right)^{\frac{m+1}{3}} C_g^{*1/3} \tag{1.30}$$

1.4.1.3 Estimation of the Non-Dimensional Flux P_{ew}

Combining Eqs. (1.15) and (1.20), the following equation of flux is obtained,

$$v_w C_g = k\left(C_g - C_0 \right) \tag{1.31}$$

The non-dimensional form of the above equation is

$$P_{ew} = Sh\left(\frac{C_g^* - 1}{C_g^*} \right) \tag{1.32}$$

Combining Eqs. (1.32) and (1.23), the length averaged permeate flux for a Newtonian fluid becomes,

$$\overline{Pe_w} = 1.816 \left(Re.Sc\frac{d}{L} \right)^{\frac{1}{3}} \left(e^{-\frac{2}{3}\alpha C_0 \left(C_g^* - 1 \right)} \right)^{0.14} \left(\frac{T_0}{T} \right)^{\frac{m+1}{3}} \left(C_g^{*1/3} - C_g^{*-2/3} \right) \tag{1.33}$$

For $C_g^* < e^3$, that is $C_g^* \ll 20$, $\left(C_g^{*1/3} - C_g^{*-2/3} \right)$, is reduced to $\ln C_g^*$, thus

$$\overline{Pe_w} = 1.816 \left(Re.Sc\frac{d}{L} \right)^{\frac{1}{3}} \left(e^{-\frac{2}{3}\alpha C_0 \left(C_g^* - 1 \right)} \right)^{0.14} \left(\frac{T_0}{T} \right)^{\frac{m+1}{3}} \ln C_g^* \tag{1.34}$$

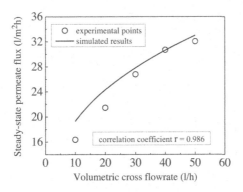

FIGURE 1.3   UF of kiwi fruit juice with PEEKWC hollow fiber membranes. Effect of feed flow rate on steady-state permeate flux (operating conditions: $\Delta P = 50$ kPa; T = 25°C) (Mondal et al., 2011b).

The viscosity of kiwi fruit juice having suspended solids of 10% was analyzed by increasing the temperature of the juice from 16 to 40°C (Mondal et al., 2011b). The value of the exponent '*m*' was obtained by curve-fitting the experimental data points, and it was found to be 0.6 ± 0.05. The viscosity of the fruit juice can be decreased by increasing the temperature and also by decreasing the suspended solids concentration.

Modified poly(ether ether ketone) (PEEKWC) hollow fiber UF membranes embedded in a 2 cm long glass module (consisting of 5 hollow fibers) were used for the filtration experiments. The total available membrane filtration area was 20 cm². The inside and outside diameter of the fiber were 1.41 and 1.96 mm, respectively.

Figure 1.3 shows the effect of the flow rate on the steady-state permeate flux at a transmembrane pressure drop of 50 kPa and a temperature of 25°C. With increase in crossflow velocity, the concentration polarization decreases due to enhanced forced convection, leading to an enhancement in mass transfer coefficient and consequently to the permeate flux. The values of the parameters D, α and C_g^* in Eq. (1.34) were estimated by minimizing the sum of squares between the calculated and experimental data of steady state permeate flux at various operating conditions. The estimated values of these three parameters were 3.6×10^{-11} m²/s, 0.007 ± 0.0008 and 4.2 ± 0.06, respectively. C_g^* is the ratio of the gel concentration to the initial bulk concentration. The value of C_g^* was found to be 4.2, which signifies that the gel layer adds a considerable resistance to the driving force. The results showed that the value of C_g^* is well below the limit of 20.08 (e^3). It is to be noted here that the Sieder-Tate viscosity correction factor, $\left(e^{-\frac{2}{3}\alpha C_0 \left(C_g^* - 1 \right)} \right)^{0.14}$ has a significant value of around 0.81. This signifies that the effect of viscosity has a strong influence on the flux decline mechanism.

In Figure 1.4, the effects of temperature on the steady-state permeate flux are described at a constant flow rate of 40 l/h and ΔP of 50 kPa. It is quite evident from the figure that on increasing the temperature, the permeate flux increases. This is because on increasing the temperature, the mass transfer coefficient k_T increases

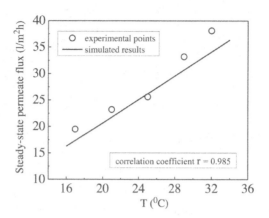

FIGURE 1.4   UF of kiwi fruit juice with PEEKWC hollow fiber membranes. Effect of temperature on steady-state permeate flux (operating conditions: $\Delta P = 50$ kPa; $Q_f = 40$ l/h) (Mondal et al., 2011b).

according to Eq. (1.30). It may be noted that the variation of mass transfer coefficient with temperature can be derived from Eq. (1.30) and the dependence of mass transfer coefficient with temperature is $k_T = k_{T_0} \left(\dfrac{T}{T_0} \right)^{\frac{2(m+1)}{3}}$.

1.4.1.4   Batch Mode of Operation

In this mode of operation, retentate stream is recycled back to the feed tank, but the permeate is continuously taken out. This results in an increase in feed concentration accompanied by a reduction in feed volume with time of operation.

Considering an overall material balance, the following equation is obtained,

$$\frac{d}{dt}\left(\rho_f V\right) = -v_w A_m \rho_p \tag{1.35}$$

where, ρ_f and ρ_p are densities in feed and permeate streams; V is the feed volume and A_m is the effective membrane area. Assuming $\rho_f \approx \rho_p$ the above equation is modified as

$$\frac{dV}{dt} = -v_w A_m \tag{1.36}$$

Using overall species balance of gel forming component, the following equation is obtained,

$$\frac{d}{dt}(C_b V) = -v_w A_m C_p \tag{1.37}$$

Since concentration of the gel forming material in the permeate is zero ($C_p = 0$) (Cheryan, 1998), the above equation reduces to a simple algebraic equation

$$C_b V = C_0 V_0 \qquad (1.38)$$

With initial condition as $V = V_0$ at $t = 0$.

Now, following the material balance for the gel-forming component in the concentration boundary layer results in the following equation (De and Bhattacharya, 1997b):

$$\text{for } 0 < y < \delta, \quad j_1 = \text{mass flux} = \rho_g \frac{dH}{dt} = v_w C_1 - D \frac{dc_1}{dy}, \qquad (1.39)$$

The pertinent boundary conditions are,

$$C_1 = C_b\,(t) \text{ at } y = 0 \qquad (1.40)$$

$$C_1 = C_g \text{ at } y = \delta \qquad (1.41)$$

The solution of the Eq. (1.37) using the above stated boundary conditions represents the variation of the gel layer thickness (H) with time,

$$\rho_g \frac{dH}{dt} = v_w \frac{C_g - C_b \exp\left(\dfrac{v_w}{k}\right)}{1 - \exp\left(\dfrac{v_w}{k}\right)} \qquad (1.42)$$

where k is the mass transfer coefficient defined as D/δ. In this case, the expression of mass transfer coefficient is not same as Eq. (1.30). This is because the boundary condition of the concentration profile within the concentration boundary layer at the edge is no longer initial feed concentration (C_0). It becomes bulk concentration that is a function of time, $C_b(t)$. The non-dimensional solute balance equation within the concentration boundary layer in this case can be written as,

$$\frac{1}{4}\frac{\partial C^*}{\partial \tau} + Ay^* \frac{\partial C^*}{\partial x^*} - \frac{Pe_w}{2}\frac{\partial C^*}{\partial y^*} = \frac{\partial^2 C^*}{\partial y^{*2}} \qquad (1.43)$$

$$(T_1) \qquad (T_2) \qquad (T_3) \qquad (T_4)$$

where the non-dimensional time is defined as $\tau = tD/d^2$. Next, an order of magnitude analysis of Eq. (1.43) is carried out term-wise. $O(x^*)$ is 1; order of y is same as that of thickness of concentration boundary layer, $\delta \approx \dfrac{D}{k} = \dfrac{10^{-11}}{10^{-6}} = 10^{-5}$.

Thus, $O(y^*)$ is $\dfrac{10^{-5}}{10^{-3}} = 10^{-2}$. $O(A)$ is $\dfrac{u_0 d^2}{DL} = \dfrac{1 \times 10^{-6}}{10^{-11} \times 10^{-1}} = 10^6$. $O(Pe_w)$ is $\dfrac{v_w d}{D} = \dfrac{10^{-6} \times 10^{-3}}{10^{-11}} = 10^2$. Therefore, the order of the terms, T_2, T_3 and T_4 is 10^4. Thus, it may be noted that T_1 has significant magnitude compared to the other three terms up to a time of operation of 100 s. Beyond 100 s, it is reduced in order of magnitude. Hence, comparing the full operation time in this experiment (460 min), T_1 is small enough to be ignored. Therefore, we can take recourse to a quasi-steady state analysis for estimation of the concentration boundary layer profile. The governing equation of solute mass balance is same as Eq. (1.6). The concentration profile can be approximated as Eq. (1.7), along with the boundary conditions Eqs. (1.8) and (1.10). Eq. (1.9) now becomes,

$$\text{at } y^* = \delta^*, \quad C = C_b; \quad C^* = C_b^* \tag{1.44}$$

The concentration profile within the boundary layer now becomes,

$$C^* = C_g^* - 2\left(C_g^* - C_b^*\right)\left(\frac{y^*}{\delta^*}\right) + \left(C_g^* - C_b^*\right)\left(\frac{y^*}{\delta^*}\right)^2 \tag{1.45}$$

Proceeding exactly like previous case as described before, the mean concentration within the boundary layer is,

$$C_m^* = \frac{1}{3}\left(C_g^* + 2C_b^*\right) \tag{1.46}$$

and,

$$\frac{\mu_m}{\mu_w} = e^{-\frac{2}{3}\alpha C_0\left(C_g^* - C_b^*\right)} \tag{1.47}$$

The Sherwood number relation is modified as,

$$\overline{Sh_L} = 1.816\left(\text{Re}.Sc\frac{d}{L}\right)^{\frac{1}{3}}\left(e^{-\frac{2}{3}\alpha C_0\left(C_g^* - C_b^*\right)}\right)^{0.14}\left(\frac{C_g^*}{C_b^*}\right)^{\frac{1}{3}} \tag{1.48}$$

Including temperature correction, the average Sherwood number is,

$$\overline{Sh_L} = 1.816\left(\text{Re}.Sc\frac{d}{L}\right)^{\frac{1}{3}}\left(e^{-\frac{2}{3}\alpha C_0\left(C_g^* - C_b^*\right)}\right)^{0.14}\left(\frac{T_0}{T}\right)^{\frac{m+1}{3}}\left(\frac{C_g^*}{C_b^*}\right)^{\frac{1}{3}} \tag{1.49}$$

The average dimensionless permeate flux becomes

$$\overline{Pe_w} = 1.816\left(\text{Re}.Sc\frac{d}{L}\right)^{\frac{1}{3}}\left(e^{-\frac{2}{3}\alpha C_0\left(C_g^* - C_b^*\right)}\right)^{0.14}\left(\frac{T_0}{T}\right)^{\frac{m+1}{3}}\left[\left(\frac{C_g^*}{C_b^*}\right)^{\frac{1}{3}} - \left(\frac{C_g^*}{C_b^*}\right)^{-\frac{2}{3}}\right] \tag{1.50}$$

For $\dfrac{C_g^*}{C_b^*} < e^3$, that is $C_g^* \ll 20C_b^*$, $\left(\dfrac{C_g^*}{C_b^*}\right)^{\frac{1}{3}} - \left(\dfrac{C_g^*}{C_b^*}\right)^{-\frac{2}{3}}$, is reduced to $\ln\left(\dfrac{C_g^*}{C_b^*}\right)$. Under

this condition, the final expression of length averaged permeate flux is,

$$\overline{Pe_w} = 1.816\left(\mathrm{Re}.Sc\frac{d}{L}\right)^{\frac{1}{3}}\left(e^{-\frac{2}{3}\alpha C_0\left(C_g^*-C_b^*\right)}\right)^{0.14}\left(\frac{T_0}{T}\right)^{\frac{m+1}{3}}\ln\left(\frac{C_g^*}{C_b^*}\right) \tag{1.51}$$

There are several instances of the rheology of fruit juice exhibiting non-Newtonian characteristics such as that of carrot (Vandresen et al., 2009), blueberry (Nindo et al., 2005), mango (Dak et al., 2007), pummel (Chin et al., 2009), pineapple (Dak et al., 2008), pomegranate (Yildiz et al., 2009), sugar cane (Filho et al., 2011), mandarin (Falguera et al., 2010), date (Gabsi et al., 2013), guava (Sánchez et al., 2009) and blood orange (Mizrahi and Berk, 1972), which confirms that the power law model is more appropriate. Considering a power law fluid rheology, the shear stress (τ) relationship with the shear rate $(\dot{\gamma})$ can be described by a simple polynomial function as

$$\tau = \beta\,\dot{\gamma}^n \tag{1.52}$$

where β is the consistency index and n is the flow behavior index. Following similar derivation for a Newtonian fluid, the Sherwood number can be expressed for the non-Newtonian fluid as

$$\overline{Sh} = 1.145\left(3+\frac{1}{n}\right)^{1/3}\left(\mathrm{Re}.Sc.\frac{d}{L}\right)^{1/3}\left(e^{-\frac{2}{3}\alpha C_0\left(C_g^*-C_b^*\right)}\right)^{0.14}\left(\frac{C_g^*}{C_b^*}\right)^{1/3} \tag{1.53}$$

Further the length averaged permeate flux becomes,

$$\overline{Pe_w} = 1.145\left(3+\frac{1}{n}\right)^{1/3}\left(\mathrm{Re}.Sc\frac{d}{L}\right)^{\frac{1}{3}}\left(e^{-\frac{2}{3}\alpha C_0\left(C_g^*-C_b^*\right)}\right)^{0.14}\left[\left(\frac{C_g^*}{C_b^*}\right)^{1/3}-\left(\frac{C_g^*}{C_b^*}\right)^{-2/3}\right] \tag{1.54}$$

The flux v_w can be expressed using the phenomenological equation,

$$v_w = \frac{\Delta P}{\mu\left(R_m + R_g\right)} \tag{1.55}$$

where, R_m is the hydraulic membrane resistance determined experimentally and R_g is the gel layer resistance. The gel layer is assumed to be a deposit of porous cake, using the filtration concept. Hence, the gel layer resistance and its characteristics are described with the platform of traditional cake filtration theory (Bhattacharjee et al., 1996). So, R_g is expressed as,

$$R_g = \psi\left(1-\varepsilon_g\right)\rho_g H \tag{1.56}$$

where ψ is the specific cake resistance, ε_g is the porosity of the cake, ρ_g is the density of the cake layer. Since ε_g, ρ_g are all constants during the experiment, the product $\psi(1-\varepsilon_g)\rho_g$ is clubbed together into a single parameter and is treated as another constant (ξ) during the course of the simulation. Thus, $R_g = \xi H$. Combining Eq. (1.42) and the relation of R_g with H, the governing equation of gel layer thickness becomes,

$$\rho_g \frac{dH}{dt} = v_w \frac{C_g - C_b \exp\left(\dfrac{\Delta P}{k\mu(R_m + R_g)}\right)}{1 - \exp\left(\dfrac{\Delta P}{k\mu(R_m + R_g)}\right)} \tag{1.57}$$

The initial condition of the above equation is

$$H = 0 \text{ at } t = 0 \tag{1.58}$$

From Eq. (1.38) the bulk concentration becomes

$$C_b = \frac{C_0 V_0}{V} \tag{1.59}$$

It may be noted that as time of operation proceeds, the effective fiber diameter decreases by deposition of gel layer, and it is quantified as,

$$d(t) = d(t = 0) - 2H(t) \tag{1.60}$$

Consequently, the crossflow velocity $u_0(t)$ inside the fiber changes as,

$$u_0(t) = \frac{4Q}{\pi[d(t)]^2} \tag{1.61}$$

The above expressions of fiber diameter and crossflow velocity within fiber have been utilized to evaluate the mass transfer coefficient in Eq. (1.49). The governing equation of volume at any time point is given by Eq. (1.36). Eqs. (1.36, 1.57, 1.59–1.61) present a system of differential-algebraic equations. These are solved numerically using the fourth-order Runge-Kutte method. Thus, the time profiles of volume reduction factor (VRF), retentate concentration, gel layer thickness, etc., are obtained.

The kiwi fruit juice, with a suspended solids concentration of 10%, was clarified according to the batch concentration configuration in optimal operating conditions ($\Delta P = 75$ kPa, T = 25°C, $Q_f = 40$ l/h) (Mondal et al., 2011b). The intrinsic membrane resistance was determined to be 1.12×10^{12} m^{-1}, while the initial feed concentration C_0 was 100 kg/m^3. Figure 1.5a shows the decline of permeate flux as well as VRF with time. The decline of flux is very rapid during the initial few minutes. Beyond 30 minutes, the flux continues to decrease gradually with time due to enhanced

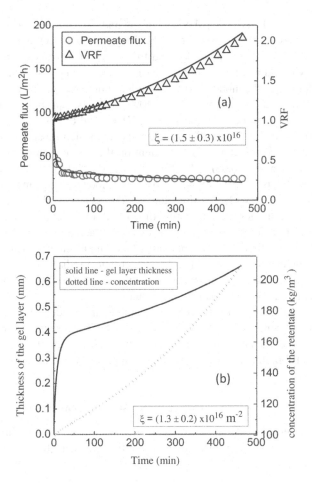

FIGURE 1.5   (a) Time course of permeate flux and VRF in the UF of kiwi fruit juice according to the batch concentration mode (operating conditions: $\Delta P = 75$ kPa; $T = 25°C$; $Q_f = 40$ l/h). (b) Evolution of the gel layer thickness and retentate concentration in the UF of kiwi fruit juice (Mondal et al., 2011b).

concentration polarization. Figure 1.5a shows remarkable agreement between the calculated and experimental observations. As the experiment progresses, the volume of the feed (or retentate) decreases, leading to an increase in VRF as shown in this figure. Thus, to maintain the mass conservation, the concentration of the feed increases with time as represented by Eq. (1.38), and this variation is illustrated in Figure 1.5b. The variation of the thickness of the gel layer with time is also presented in this figure. The gel layer thickness increases with time non-linearly, and it goes up to 0.6 mm at the end of the experiment. It may be mentioned here that the Sherwood number decreases from 307 to 258 during the entire course of the experiment.

The model proposed by Mondal et al. (2016), developed from the first principle boundary layer, described the mass transport phenomena involved in the clarification of blood orange juice by using different hollow fiber membranes (polysulphone [PS] with molecular weight cut-off [MWCO] of 50 and 10 kDa and polyacrylonitrile [PAN] with MWCO of 50 kDa). The rheology of the blood orange juice was described using the power law model, and the values of the consistency index (β) and flow behavior index (n) were found to be 1.766 Pa.s^{2-n} and 0.93. The gel-forming solute (C as defined in the model) is the suspended solid content. The feed-suspended solid concentration (C_0) was found to be 10% w/w. The optimized values of the parameters (intrinsic property of the solution) D, C_g^* and α were (4.9 ± 0.1) × 10^{-11} m^2/s, 6.1 ± 0.08 and (5 ± 0.2) × 10^{-4} m^3/kg, respectively. However, the specific cake resistance (clubbed parameter ξ) is dependent on the solute-membrane system and was calculated as (6.57 ± 0.08) × 10^{-11} m^{-2} for the PS 100 kDa membranes. There was variation of the permeate flux with crossflow velocity for this membrane, as the permeate flux increased with crossflow rate. This is because with crossflow velocity, the forced convection increases, arresting the growth of the concentration polarization layer. The deviation of the predicted result with actual experimental data was less than ±7%. In the first few moments, other blocking mechanisms are also prominent, causing the inaccuracy of the present model in its predictions. Moreover, at the start of the experiment, the pump takes some time to get the flow rate to be steady, as it adds perturbed unsteady loads to the system.

There was a VRF during the batch filtration mode. It is evident that the theoretical prediction was within ±10% of the experimental data. The strength in model results is exhibited by extrapolating the VRF values beyond the actual filtration time. In fact, this is useful in understanding the long-term performance output of the system.

1.4.2 Fruit Juice Filtration Model in Turbulent Flow Systems

The fully developed, steady-state turbulent velocity profile for flow through closed conduit can be expressed in terms of the coordinate system fixed at the walls as (Reichardt, 1951),

$$u^+ = \frac{1}{\lambda}\ln\left(1 + \lambda y^+\right) + 7.8\left[1 - \exp\left(-\frac{y^+}{11}\right) - \frac{y^+}{11}\exp\left(-0.33y^+\right)\right] \quad (1.62)$$

where λ is the von-Karman constant equal to 0.41 (Bird et al., 2007); y^+ is non-dimensional transverse dimension $\left(= \frac{u*}{\mu/\rho}y\right)$; $u^+ = u/u*$; $u*$ is the friction velocity $\left(= \sqrt{\tau_w/\rho}\right)$ and the wall shear stress $\tau_w = 0.03954\rho u_0^2\left(\frac{\mu}{\rho u_0 d}\right)^{0.25}$ (Fox and McDonald, 2004). The velocity profile in Eq. (1.64) is applicable for both the viscous sub layer and buffer zone of the turbulent flow.

Similar to the case of the laminar flow regime, the species transport equation (in non-dimensional terms) can be expressed as

$$\text{Re}\,Sc\frac{d}{L}U\frac{\partial C^*}{\partial X} - Pe_w\frac{\partial C^*}{\partial Y} = \frac{\partial^2 C^*}{\partial Y^2} \quad (1.63)$$

where the velocity profile $U = \dfrac{u}{u_0} = P\,ln[1+QY] + R[1-\exp(-SY) - SY\exp(-TY)]$.
The non-dimensional terms involved in the expression of U are $P = 0.5Re^{-0.125}$;
$Q = 0.082Re^{0.875}$; $R = 1.56Re^{-0.125}$; $S = 0.018Re^{0.875}$ and $T = 0.066Re^{0.875}$.

Following an integral method of analysis considering a quadratic profile of the concentration as explained for the laminar situation, the expression for the length averaged Sherwood number can be obtained as

$$\overline{Sh_L} = \int_0^1 Sh(X)dX = 2\int_0^1 \frac{dX}{\Delta} \tag{1.64}$$

where the non-dimensional mass transfer boundary layer thickness can be computed from

$$\mathrm{Re}\,Sc\frac{d}{L}\Delta\frac{d\Delta}{dX}I_1 = \frac{1}{C_g^*} \tag{1.65}$$

where $I_1 = \displaystyle\int_0^\Delta U\left(\frac{Y}{\Delta^2} - \frac{Y^2}{\Delta^3}\right)dY$

$$= \int_0^\Delta \left\{P\ln[1+QY] + R[1-\exp(-SY) - SY\exp(-TY)]\right\}\left(\frac{Y}{\Delta^2} - \frac{Y^2}{\Delta^3}\right)dY$$

The rest of the analysis for estimating the permeate flux in the total recycle and batch mode of operation is similar to the case of laminar flow. A mathematical model describing the permeate flux decline in the clarification of fruit juices with hollow fiber membranes under turbulent flow was presented by Mondal et al. (2014). The developed model was used to explain the pomegranate juice clarification using PEEKWC and PS hollow fiber membrane modules with a total membrane surface area of 26.9 cm^2 and 32.3 cm^2, respectively. The membrane specifications are stated in Table 1.1.

TABLE 1.1

Characteristics of Hollow Fiber Membranes Used in the Clarification of Pomegranate Juice

Membrane Material	PEEKWC	PS
Internal diameter (mm)	1.19	1.43
External diameter (mm)	1.62	1.73
Thickness (mm)	0.215	0.15
Hydraulic permeability (L/m^{2}hbar)	536.14	654.81

The effect of the Reynolds number on the profiles of the mass transfer boundary layer and Sherwood number for the PEEW-WC membrane is presented in Figure 1.6.

From Figure 1.6, it is evident that the mass transfer boundary layer is in the developing state for this length of the module. Hence, the application of the conventional heat and mass transfer analogies, which considers fully developed boundary layer, for prediction of the Sherwood number, would seriously undermine the permeate flux in this case. The thickness of the mass transfer boundary layer is affected by the turbulence in the flow channel. It decreases with enhanced turbulence. As the Reynolds number increases, the forced convection restricts the growth of the mass transfer boundary layer. For example, at the end of the hollow fiber, the thickness of the mass transfer boundary layer decreases from 7 to 4.5 mm as Re increases

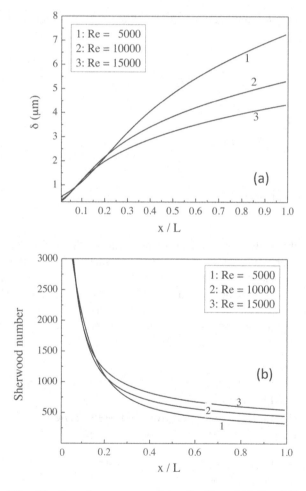

FIGURE 1.6 Microfiltration of pomegranate juice. Profiles of (a) δ and (b) the Sherwood number with Re, in total recycle mode, using a PEEKWC membrane. The values of the physical parameters are $\rho_g = 1400 \text{ kg/m}^3$; $\varepsilon_g = 0.3$; $D = 5 \times 10^{-11} \text{ m}^2/\text{s}$; $C_g^* = 3.2$; $\xi = 1.5 \times 10^{17} \text{ m}^{-2}$ (Mondal et al., 2014).

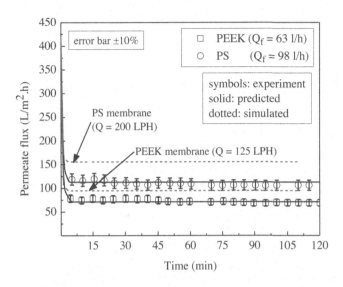

FIGURE 1.7 Comparison of the permeate flux profile with experimental and theoretical results. The dotted lines are for simulated results of two different crossflow velocities. The optimized values of the physical parameters are $\varepsilon_g = 0.3$; $\rho_g = 1400$ kg/m^3; $\xi = 1.5 \times 10^{17}$ m^{-2}; $D = 2.2 \times 10^{-12}$ m^2/s and $C_g^* = 3.2$ for PEEKWC hollow fiber membranes [4.6 in case of PS membrane] (Mondal et al., 2014).

from 5000 to 15000. The effect of the Re number on the Sherwood number profile is shown in Figure 1.6b. It is apparent from this figure that Sherwood number profiles have inverse trends of thickness of mass transfer boundary layers, confirming the theory. The Sherwood number (mass transfer co-efficient) decreases sharply up to $x/L = 0.2$ due to the rapid growth of mass transfer boundary layer and stabilizes thereafter. The Sherwood number increases with the Reynolds number, indicating a decrease in thickness of the mass transfer boundary layer with the Reynolds number, resulting in less resistance offered to solvent flux across the membrane.

The validation of the developed model is illustrated in Figure 1.7 by using the optimized values of the physical parameters. The optimized set of parameters is $\varepsilon_g = 0.3$; $\rho_g = 1400$ kg/m^3; $\xi = 1.5 \times 10^{17}$ m^{-2}; $D = 2.2 \times 10^{-12}$ m^2/s and $C_g^* = 3.2$ (4.6 in the case of the PS membrane).

The model results show that the permeate flux values are well within ±10% of the experimental data. Two additional set of simulated flux profiles corresponding to twice the crossflow velocity as used in the experiment were also presented in the figure (dotted lines). It is understood from this figure that gel layer resistance is a major contribution to the flux decline behavior. The steady state flux is reduced to 25% of its pure water flux.

In the case of the batch mode of filtration, the comparison of the predicted model results with experimental data is presented in Figure 1.8. The values of the model parameters, as obtained by optimizing the flux profiles in case of total recycle mode of operation, are used to predict the performance of batch modes of operation. Experiments were conducted up to 4 hours in case of the PS membrane. However, the

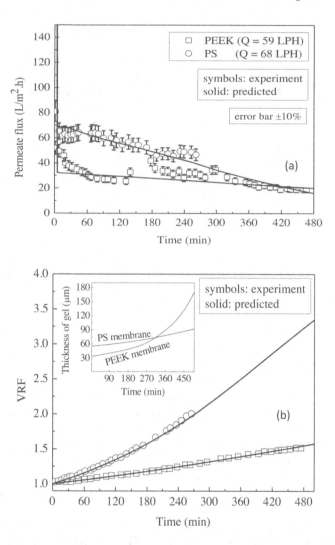

FIGURE 1.8 (a) Comparison of the experimental and predicted flux profiles for batch mode filtration (the values of the physical parameters are those optimized in total recycle mode). (b) Comparison of the experimental and predicted VRF profiles for different membrane systems; inset: gel layer dynamics for this operating condition corresponding to the experimental results (the values of the physical parameters are those optimized in total recycle mode) (Mondal et al., 2014).

model results are extended for 8 hours. It can be seen that for most of the data points, the model results are within ±10% of the experimental results. In the case of profiles for the VRF, the model prediction is almost equal to the experimental results. The results of batch modes suggest that for different operating conditions, the permeate flux approaches a limiting value after a considerable time of operation. Similar to the total recycle mode operation, the flux reduces to almost 10% of the pure water flux.

The transient profile of the gel layer thickness is obtained from the simulation as shown in Figure 1.6b. This suggests that in the case of the PEEKWC membrane, the development of the gel layer is sharper compared to the PS membrane.

1.4.2.1 Conclusion

A mathematical model has been developed to describe the physical mass transport phenomena and quantify the various extents of fouling in the clarification of fruit juices with hollow fiber membranes using different membrane materials and operating conditions. The model results have been validated in close agreement with the experimental data. The generalized model formulated in this work will be useful in predicting and analyzing the flux profile for micro- or ultra-filtration of various fruit juice mixtures using different polymeric membranes.

A further improvement on the predictive ability of the present model can be achieved by including the viscoelastic behavior in the fluid rheology (as many fruit juices exhibit such rheological properties), the concentration dependence on mixture diffusivity of the polydisperse solution and particle-wall interactions due to adsorption and sticking of solute particles on the membrane surface.

1.5 NOMENCLATURE

A	product of Reynolds number, Schmidt number and d/L, $u_0 d^2/DL$
A_m	effective area of the membrane surface, m^2
C	concentration, kg/m^3
C_0	initial concentration, kg/m^3
C_b	bulk concentration, kg/m^3
C_g	gel concentration, kg/m^3
C_p	permeate concentration, kg/m^3
C^*	non-dimensional concentration
C_g^*	non-dimensional gel concentration
d	membrane pore diameter, as a function of time, m
D	effective diffusivity, m^2/s
H	thickness of the gel layer, m
k	mass transfer coefficient, m/s
P_{ew}	Non-dimensional flux
Q	Volumetric flow rate, m^3/s
r	local radius of the tube, m
R	radius of the tube, m
R_g	resistance of the gel layer, m^{-1}
R_m	hydraulic resistance of the membrane, m^{-1}
Re	Reynolds number
Sc	Schmidt number
Sh	Local Sherwood number
Sh_L	average Sherwood number, averaged along the length L
T	temperature of the feed, °C
T_0	reference standard temperature, °C
u	velocity in x – direction, m/s

u_0 maximum radial velocity, m/s
v velocity in y – direction, m/s
V feed volume, m^3
V_0 initial feed volume, m^3
v_w permeate flux, m^3/m^2s
x axial coordinate, m
y transverse coordinate, m
x^* non-dimensional axial distance
y^* non-dimensional transverse distance

1.5.1 GREEK SYMBOLS

α an adjustable parameter in Sieder – Tate correction factor, m^3/kg
β product of α and C_0
ΔP transmembrane pressure, kPa
δ^* non – dimensional thickness of the boundary layer
ε_g porosity of the gel layer
μ viscosity of the solution, mPa.s
μ_m viscosity of the solution at mean concentration, mPa.s
μ_0 viscosity of the pure liquid, i.e. without any suspended solids, mPa.s
μ_w viscosity of the solution near the wall, mPa.s
ρ_f density of the feed solution, m^3/kg
ρ_g density of the gel layer, m^3/kg
ρ_p density of the permeate, m^3/kg
ξ product of $\psi(1 - \varepsilon_g)\rho_g$ (m^{-2})
ψ specific cake resistance (m^{-1})

REFERENCES

Barros, de S.T.D., Andrade, C.M.G., Mendes, E.S., Peres, L. (2003). Study of fouling mechanism in pineapple juice clarification by ultrafiltration, Journal Membrane Science, 215, 213–224.

Belfort, G., Davis, R.H., Zydney, A.L. (1994). The behavior of suspensions and macromolecular solutions in crossflow microfiltration, Journal of Membrane Science, 96, 1–58.

Bhattacharjee, C., Saxena, V.K., Dutta, S. (2017). Fruit juice processing using membrane technology: A review, Innovative Food Science & Emerging Technologies, 43, 136–153.

Bhattacharjee, S., Sharma, A., Bhattacharya, P.K. (1996). A unified model for flux prediction during batch cell ultrafiltration, Journal of Membrane Science, 111, 243–258.

Bird, R.B., Stewart, W.E., Lightfoot, E.N. (2007). Transport Phenomena (2nd ed.). New York: John Wiley & Sons.

Blatt, W.F., Dravid, A., Michaels, A.S., Nelson, L. (1970). Solute polarization and cake formation in membrane ultrafiltration: Causes, consequences and control techniques, In: Flinn, J.E. (Ed.), Membrane Science & Technology (pp. 47–75). New York: Plenum Press.

Bouchard, C.R., Carreau, P.J., Matsuuara, T., Sourirajan, S. (1994). Modeling of ultrafiltration: Prediction of concentration polarization effects, Journal of Membrane Science, 97, 215–229.

Bowen, W.R., Williams, P.M. (2001). Prediction of the rate of cross-flow ultrafiltration of colloids with concentration-dependent diffusion coefficient and viscosity-theory and experiment, Chemical Engineering Science, 56, 3083–3099.

Braddock, R.J., Goodrich, R.M. (2003). Processing technologies to enhance fresh flavor of citrus juice, In: Cadwallader, K.R., Weenen, H. (Eds.), Freshness and Shelf Life of Foods. ACS Symposium Series, Vol. 836, pp. 292–301). Washington: American Chemical Society.

Cassano, A., Donato, L., Drioli, E. (2007). Ultrafiltration of kiwifruit juice: Operating parameters. Juice quality and membrane fouling, Journal of Food Engineering, 79, 613–621.

Chen, J.C., Li, Q., Elimelech, M. (2004). In situ monitoring techniques for concentration polarization and fouling phenomena in membrane filtration, Advances in Colloid and Interface Science, 107, 83–108.

Cheryan, M. (1998). Ultrafiltration and Microfiltration Handbook. Lancaster: Technomic Publishing Company.

Chin, N.L., Chan, S.M., Yusof, Y.A., Chuah, T.G., Talib, R.A. (2009). Modeling of rheological behavior of pummelo juice concentrates using master-curve, Journal of Food Engineering, 93, 134–140.

Cussler, E.L. (1998). Diffusion: Mass Transfer in Fluid System. Cambridge, UK: Cambridge University Press.

Dak, M., Verma, R.C., Jaaffrey, S.N.A. (2007). Effect of temperature and concentration on Rheological properties of "Kesar" mango juice, Journal of Food Engineering, 80, 1011–1015.

Dak, M., Verma, R.C., Jain, M.K. (2008). Mathematical models for prediction of rheological parameters of pineapple juice. International Journal of Food Engineering, 4 3(3).

DasGupta, S., Sarkar, B. (2012). Membrane applications in fruit processing technologies, In: Rodrigues, S., Fernandes, F.A.N. (Eds.), Advances in Fruit Processing Technologies (pp. 87–148). Boca Raton, FL: CRC Press.

Davis, R.H. (1992). Modeling of fouling of crossflow microfiltration membranes, Separation & Purification Reviews, 21, 75–126.

De, S., Bhattacharya, P.K. (1997a). Prediction of mass transfer coefficient with suction in the applications of reverse osmosis and ultrafiltration, Journal of Membrane Science, 128, 119–131.

De, S., Bhattacharya, P.K. (1997b). Modeling of ultrafiltration process for a two-component aqueous solution of low and high (gel-forming) molecular weight solute, Journal of Membrane Science, 136, 57–69.

De, S., Bhattacharya, S., Sharma, A., Bhattacharya, P.K. (1997). Generalized integral and similarity solutions for concentration profiles for osmotic pressure controlled ultrafiltration, Journal of Membrane Science, 130, 99–121.

De, S., Bhattacharya, P.K. (1999). Mass transfer coefficient with suction including property variations in applications of cross-flow ultrafiltration, Separation and Purification Technology, 16, 61–73.

Denisov, G.A. (1999). Theory of concentration polarization in cross-flow ultrafiltration: Gel-layer model and osmotic-pressure model, Journal of Membrane Science, 91, 173–187.

Falguera, V., Ruiz, J.F.V., Alins, V., Ibarz, A. (2010). Rheological behaviour of concentrated mandarin juice at low temperatures, International Journal of Food Science and Technology, 45, 2194–2200.

Field, R., Aimar, P. (1993). Ideal limiting fluxes in ultrafiltration: Comparison of various theoretical relationships, Journal of Membrane Science, 80, 107–115.

Field, R.W., Wu, D., Howell, J.A., Gupta, B.B. (1995). Critical flux concept for microfiltration fouling, Journal of Membrane Science, 100, 259–272.

Filho, Z.A., Telis, V.R.N., de Oliveira, E.B., Coimbra, J.S.R., Romero, J.T. (2011). Rheology and fluid dynamics properties of sugarcane juice, Biochemical Engineering Journal, 53, 260–265.

Fox, R.W., McDonald, A.T. (2004). Introduction to Fluid Mechanics (5th ed.). New Delhi: Wiley-India.

Gabsi, K., Trigui, M., Barrington, S., Helal, A.N., Taherian, A.R. (2013). Evaluation of rheo-logical properties of date syrup, Journal of Food Engineering, 117, 165–172.

Gill, W.N., Wiley, D.E., Fell, C.J.D., Fane, A.G. (1988). Effect of viscosity on concentration polarization in ultrafiltration, AIChE Journal, 34, 1563–1567.

Girard, B., Fukumoto, L.R. (2000). Membrane processing of fruit juices and beverages: A review, Critical Reviews in Food Science and Nutrition, 40, 91–157.

Guell, C., Ferrando, M., Lopez, F. (Eds.) (2009). Monitoring and Visualizing Membrane-Based Processes. Weinheim: Wiley-VCH Verlag GmbH.

Happel, J., Brenner, H. (1965). Low Reynolds Number Hydrodynamics. New Jersey: Prentice Hall.

Hermia, J. (1982). Constant pressure blocking filtration laws: Applications to power-law non-Newtonian fluids, Transactions of the Institution of Chemical Engineers, 60, 183–187.

Hojjatpanah, G., Emam-Djomeh, Z., Kalbasi Ashtari, A., Mirsaeedghazi, H., Omid, M. (2011). Evaluation of the fouling phenomenon in the membrane clarification of black mulberry juice, International Journal of Food Science and Technology, 46, 1538–1544.

Incropera, F.P., DeWitt, D.P. (1996). Fundamentals of Heat and Mass Transfer. Singapore: John Wiley & Sons.

Johnston, S.T., Deen, W.M. (1999). Hindered convection of proteins in agarose gels, Journal of Membrane Science, 153, 271–279.

Karode, S.K. (2001). Unsteady state flux response: A method to determine the nature of the solute and gel layer in membrane filtration, Journal of Membrane Science, 188, 9–20.

Kaur, C., Kapoor, H.C. (2001). Antioxidants in fruits and vegetables - The millennium's health, International Journal of Food Science and Technology, 36, 703–725.

Kleinstreuer, C., Paller, M.S. (1983). Laminar dilute suspension flows in plate and frame ultrafiltration units, AIChE Journal, 29, 533–539.

Kozinski, A.A., Lightfoot, E.N. (1971). Ultrafiltration of proteins in stagnation flow, AIChE Journal, 17, 81–85.

Lévêque, A. (1928). Les Lois de la Transmission de Chaleur par Convection, Annuals Mines, 13, 201–299.

Madireddi, K., Babcock, R.B., Levine, B., Kim, J.H., Stenstrom, M.K. (1999). An unsteady-state model to predict concentration polarization in commercial spiral wound mem-branes, Journal of Membrane Science, 157, 13–34.

Mizrahi, S., Berk, Z. (1972). Flow behaviour of concentrated orange juice: Mathematical treatment, Journal of Texture Studies, 3, 69–79.

Mohammad, A.W., Ng, C.Y., Lim, Y.P., Ng, G.H. (2012). Ultrafiltration in food processing industry: Review on application, membrane fouling and fouling control, Food and Bioprocess Technology, 5, 1143–1156.

Mondal, S., Cassano, A., Conidi, C., De, S. (2016). Modeling of gel layer transport during ultrafiltration of fruit juice with non-Newtonian fluid rheology, Food and Bioproducts Processing, 100, 72–84.

Mondal, S., Cassano, A., Tasselli, F., De, S. (2011b). A generalized model for clarification of fruit juice during ultrafiltration under total recycle and batch mode, Journal of Membrane Science, 366, 295–303.

Mondal, S., Chhaya, De, S. (2012). Prediction of ultrafiltration performance during clarifica-tion of stevia extract, Journal of Membrane Science, 396, 138–148.

Mondal, S., De, S., Cassano, A., Tasselli, F. (2014). Modeling of turbulent cross flow micro-filtration of pomegranate juice using hollow fiber membranes, AIChe Journal, 60, 4279–4291.

Mondal, S., Rai, C., De, S. (2011a). Identification of fouling mechanism during ultrafiltration of stevia extract, Food and Bioprocess Technology, 6, 1–10.

Nindo, C.I., Tang, J., Powers, J.R., Singh, P. (2005). Viscosity of blueberry and raspberry juices for processing applications, Journal of Food Engineering, 69, 343–350.

Nunes, S.P., Peinemann, K.V. (2001). Membrane Technology in the Chemical Industry. Weinheim: Wiley/VCH.

Opong, W.S., Zydney, A.L. (1991). Diffusive and convective protein transport through asymmetric membranes, AIChE Journal, 37, 1497–1510.

Porter, M.C. (1972). Concentration polarization with membrane ultrafiltration, Industrial & Engineering Chemistry Product Research and Development, 11, 234–248.

Probstein, R.F., Shen, J.S., Leung, W.F. (1978). Ultrafiltration of macromolecular solutions at high polarization in laminar channel flow, Desalination, 24, 1–16.

Rai, P., Majumdar, G.C., DasGupta, S., De, S. (2007). Effect of various pre treatment methods on permeate flux and quality during ultrafiltration of mosambi juice, Journal of Food Engineering, 78, 561–568.

Rai, C., Rai, P., Majumdar, G.C., De, S., DasGupta, S. (2010). Mechanism of permeate flux decline during microfiltration of watermelon (*Citrullus lanatus*) juice, Food and Bioprocess Technology, 3, 545–553.

Rai, P., Rai, C., Majumdar, G.C., DasGupta, S., De, S. (2006). Resistance in series model for ultrafiltration of mosambi (*Citrus sinensis (L.) Osbeck*) juice in a stirred continuous mode, Journal of Membrane Science, 283, 116–122.

Reichardt, H. (1951). Vollständige darstellung der turbulenten geschwindigkeitsverteilung in glatten leitungen, Journal of Applied Mathematical Mechanics, 31, 208–219.

Sablani, S.S., Goosen, M.F.A., Belushi, R.A., Wilf, M. (2001). Concentration polarization in ultrafiltration and reverse osmosis: A critical review, Desalination, 141, 269–289.

Sánchez, C., Blanco, D., Oria, R., Gimeno, A. (2009). White guava fruit and purees: Textural and rheological properties and effect of the temperature, Journal of Texture Studies, 40, 334–345.

Sarkar, B., DasGupta, S., De, S. (2008). Cross-flow electro-ultrafiltration of mosambi (*Citrus sinensis (L.) Osbeck*) juice, Journal of Food Engineering, 89, 241–245.

Shen, J.J.S., Probstein, R.F. (1977). On the prediction of limiting flux in laminar ultrafiltration of macromolecular solutions, Industrial & Engineering Chemistry Fundamentals, 16, 459–465.

Sherwood, T.K., Brian, P.L.T., Fisher, R.E., Dresner, L. (1965). Salt concentration at phase boundaries in desalination by reverse osmosis, Industrial & Engineering Chemistry Fundamentals, 4, 113–118.

Song, L., Elimelech, M. (1995). Theory of concentration polarization in crossflow filtration, Journal of Chemical Society Faraday Transactions, 91, 3389–3398.

van den Berg, G.B., Rácz, I.G., Smolders, C.A. (1989). Mass transfer coefficients in cross-flow ultrafiltration, Journal of Membrane Science, 47, 25–51.

Vandresen, S., Quadri, M.G.N., de Souza, J.A.R., Hotza, D. (2009). Temperature effect on the rheological behavior of carrot juices, Journal Food Engineering, 92, 269–274.

Vladisavljević, G.T., Vukosavljević, P., Bukvić, B. (2003). Permeate flux and fouling resistance in ultrafiltration of depectinized apple juice using ceramic membranes, Journal of Food Engineering, 60, 241–247.

Yildiz, H., Bozkurt, H., Icier, F. (2009). Ohmic and conventional heating of pomegranate juice: Effects on rheology, color and total phenolics, Food Science and Technology International, 15, 503–512.

Zydney, A.L. (1997). Stagnant film model for concentration polarization in membrane systems, Journal of Membrane Science, 130, 275–281.

2 Modelling and Sensitivity Analysis of a Fixed-Bed Reactor during the Conversion of Waste Cooking Oil to Biodiesel over Sodium Silicate Catalyst

Michael O. Daramola and Omotola Babajide

2.1 INTRODUCTION

South Africa recorded increased global energy consumption at 1.9%, driving urgent possible alternative energy sources [1]. Biodiesel has been identified as a renewable energy fuel and offers the advantages of being biodegradable, non-toxic and environmentally beneficial. Transesterification reactions for biodiesel production using alcohol and oil are facilitated by catalysts: homogeneous catalysts and heterogeneous catalysts [2]. The distinctions between these catalysts lie on the basis of their phase; homogeneous catalysts are present in the same phase as reactants and products, whereas heterogeneous catalysts are in different phases with the reactants and products [3]. Conventional methods for biodiesel production employ homogeneous base catalysts such as NaOH and KOH for their benefits, such as the ability to operate under moderate conditions with high catalyst reactivity and efficient heat transfer. The modelling of biodiesel production could be a useful tool in gaining further understanding of the transesterification process. A review of literature has shown current information on batch, semi-batch and continuously stirred tank reactors [4–6], while little or no information exists on fixed-bed reactors (FBRs), hence the need for investigative studies in modelling biodiesel production using FBRs. Mathematical models for packed bed reactors are essential for the description of the dynamic and steady-state behaviour of the system. Understanding the behaviour of the system will thus enable a better understanding of the physical and chemical phenomena occurring in the process. The developed mathematical models are needed for the description of static and dynamic behaviour for process design, optimization

DOI: 10.1201/9781003435181-2

and control by simulating the dynamic behaviour of the system [7]. FBRs are among the most widely used reactors for large-scale processing in the petroleum industry for processes such as catalytic reforming and hydro-treatment [8]. Assumptions about the behaviour of the process along with some of its parameters have to be made to simplify the mathematical models. These assumptions may not hold in reality, thus causing deviations between the mathematical model and real-life systems. In an FBR, when the reactants pass over the catalyst, they diffuse, adsorb and react on the catalyst-active sites and then desorb and diffuse back into the bulk phase. It is therefore safe to assume that convection is the dominant mechanism of heat and mass transfer. Likewise, in this study, heat transfer was considered as the system was assumed to operate isothermally. In the same vein, Likozar [4] states that for most models that describe the production of biodiesel in reactors, the mass transfer is often not accounted for. In this study, mass transfer will not be considered, as the model will be limited to reaction kinetics via the interaction of the fluid with the catalyst. The aim of this study was to develop a mathematical model and validate the results of this model using experimental data. The understanding of how the system responds to disturbances allows for system control essentially for industrial safety protocols; therefore, the response of the system to some perturbation was considered in this chapter.

2.2 METHODOLOGY

2.2.1 MODEL DEVELOPMENT

The overall mathematical model consists of the kinetic model and the reactor model. Presently, literature has yet to show the kinetics of sodium silicate catalyst to produce biodiesel using an FBR, so sodium silicate catalyst kinetics were obtained from experimental results for the purpose of this study.

The following assumptions were made during model development to simplify the mathematical model equations used in this study:

- The reactor was assumed to operate under isobaric and isothermal conditions.
- Steady-state operation was also assumed.
- The thickness of the reactor used was assumed to be so small that the temperature of the water in which the reactor was immersed equals the temperature in the reactor.
- Densities of the reactants and products are considered constants.

2.2.2 REACTION MODEL

A reaction model was developed based on the reaction kinetics for the conversion of oil and alcohol into fatty acid methyl esther (FAME). The transesterification reaction involves three stepwise reactions, namely the conversion of triglyceride to diglyceride (DG), the conversion of DG to monoglyceride (MG) and then the conversion of MG to glycerol (GL). Thus, the overall rate of reaction can be represented as follows:

Overall reaction equation:

$$TG + 3R'OH \leftrightarrow 3R'COOR + GL \tag{2.1}$$

Stepwise reactions:

$$TG + R'OH \overset{k_1}{\underset{k_{-1}}{\leftrightarrow}} DG + R'COOR \tag{2.2}$$

$$DG + R'OH \overset{k_2}{\underset{k_{-2}}{\leftrightarrow}} MG + R'COOR \tag{2.3}$$

$$MG + R'OH \overset{k_3}{\underset{k_{-3}}{\leftrightarrow}} GL + R'COOR \tag{2.4}$$

Assuming the overall rate of reaction can be represented as follows:

$$r_i = -k_f C_{TG} C_{OH} + k_r C_{COOR} C_{GL} \tag{2.5}$$

$$C_i = \frac{Q_i}{\Sigma_{i=1}^{6} V_i} \tag{2.6}$$

where C_i is the concentration of component i, Q_i is the molar flow rate of component i and V is the volume of the mixture. The reaction rate constant is independent of the concentrations of species involved in the reaction; it depends on temperature and the presence of the catalyst. In liquid systems, k can also be a function of other parameters such as ionic strength and choice of solvent; however, the effect of these parameters is small compared to that of temperature [9].

2.2.2.1 Reactor Model

The reactor model presented in this report was obtained from the modification of the model presented by Daramola et al. [9]. The mass balance for component i was performed for the FBR using the schematic presented in Figure 2.1. The solution to the model equation was obtained using the constant parameters shown in Table 2.1 and in the MATLAB environment.

$$\frac{dC_i}{dl} = r_i \pi R_1^2 \rho_{cat} \tag{2.7}$$

The following boundary conditions were used to solve the model equations for concentration profiles of the substances involved in the reactions:

- At $l = 0$, $C_i(0) = C_{i0}$
- At $l = L$, $C_i(L) = C_{iL}$

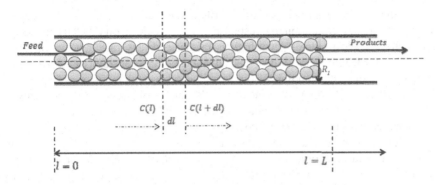

FIGURE 2.1 Schematic illustration of FBR showing the cross-sectional view with packed catalyst particle.

TABLE 2.1
Constant Parameters for the Reactor and Reaction Modelling

Parameter	Symbol	Value
Universal gas constant (J/mol/K)	R	8.314
Operating pressure (Pa)	P	101325
Density of the catalyst (kg/m³)	ρ_{cat}	240
Length of FBR (m)	L	0.14
Radius of FBR (m)	R_1	$8. \times 10^{-3}$

2.2.2.2 Sensitivity Analysis

The performance of a heterogeneous reactor depends on some parameters such as reactor design parameters (reactor size [tube diameter], length, etc.) and operating variables (feed flow rate, mass of catalyst, catalyst packing, reaction temperature and operating pressure) [9]. Reactor design involves selecting the reactor configuration/geometry, sizing the reactor, selecting and optimizing reactor operating conditions, optimizing the reactor performance, selecting suitable materials for construction, costing and scaling-up [10]. Often, the reactor design is based on some reactor design equations that provide relationships relating the reactor operation with reactor properties [11]. Reactor sizing basically involves specifying the reactor geometry, such as vessel size and type, which depends on reactor diameter and length (for tubular) or reactor diameter and height (for continuous stirred tank reactors [CSTR] or batch reactors). During reactor sizing, a designer has control over these reactor dimensions, although, specifying the dimensions depends largely on the reactor operating parameters such as the rate equation, feed flow rates, amount of catalyst and catalyst packing [11]. Furthermore, during the operation of an FBR, operating variables such as reaction temperature, operating pressure, and feed flow rate can be manipulated to optimise the reactor performance. In this chapter, the sensitivity of FBR performance to changes in properties describing the reactor geometry is presented. In the performance evaluation, the response of waste cooking oil (WCO) conversion and biodiesel yield to small perturbations in the feed flow rate, reactor diameter and the reactor length were studied. The advantage of this sensitivity analysis is not limited to better understanding of the fundamental behaviour of the FBR during transesterification of WCO to biodiesel over the solid sodium metassilicate catalyst, but it also provides further information on the optimization approach for the reactor design and perhaps the operational condition.

Sensitivity analysis is a general concept that aims at quantifying the variations of an output parameter of a system with respect to changes imposed on some input parameters [12]. Sensitivity analysis is therefore an excellent technique to help in understanding and preparing for "what to do" and "how to do" regarding the optimization of the FBR system to produce biodiesel.

For instance, if ϕ is the output function (such as waste cooking oil conversion, methanol conversion and biodiesel yield) such that ϕ depends on n input parameters δ_i (such as feed flowrate, reactor diameter, reactor length; $i = 1, 2,... n$):

$$\phi = \phi(\delta_1, \delta_2,...,\delta_n) \tag{2.8}$$

The differential form of Eq. (2.8) using the chain rule of differentiation yields:

$$d\phi = \frac{\partial\phi}{\partial\delta_1}d\delta_1 + \frac{\partial\phi}{\partial\delta_2}d\delta_2 + \cdots + \frac{\partial\phi}{\partial\delta_n}d\delta_n \tag{2.9}$$

The gradient for the parameter δ_1 then follows as:

$$\frac{d\phi}{d\delta_1} = \frac{\partial\phi}{\partial\delta_1} + \frac{\partial\phi}{\partial\delta_2}*\frac{d\delta_2}{d\delta_1} + \cdots + \frac{\partial\phi}{\partial\delta_n}*\frac{d\delta_n}{d\delta_1} \tag{2.10}$$

If it is assumed that the parameters δ_i are mutually independent of one another then:

$$\frac{d\delta_2}{d\delta_1} = \frac{d\delta_3}{d\delta_1} = \cdots = \frac{d\delta_n}{d\delta_1} = 0 \tag{2.11}$$

Therefore, the derivative of ϕ with respect to δ_1 can be approximated as:

$$\frac{d\phi}{d\delta_1} = \frac{\partial\phi}{\partial\delta_1} \approx \frac{\Delta\phi}{\Delta\delta_1} \tag{2.12}$$

Therefore, the partial derivative, $\frac{\partial\phi}{\partial\delta_1}$, is the sensitivity coefficient of the function ϕ for the input parameter δ_1. If the function ϕ is not linear with respect to parameter δ_1, $\frac{\partial\phi}{\partial\delta_1}$ will vary from point to point [13]. To obtain the sensitivity coefficients of the parameters with respect to the input variables as considered in this study, Eq. (2.12) was used and the sensitivity analysis was based on $\pm25\%$ changes in the WCO feed flow rate, reactor diameter and reactor length.

2.2.2.3 Estimation of Kinetic Parameters from Experimental Results

The graphical method was used to determine the kinetic parameters using Eqs. (2.13)–(2.16). First-order reaction rate was assumed for the transesterification reaction.

$$\frac{dC_{TG}}{dl} = -kC_{TG} \tag{2.13}$$

$$\ln[C_{TG(C)}] = -kt + \ln[C_{TG(0)}] \tag{2.14}$$

To estimate activation energy, E_A, Arrhenius equation was used as shown below:

$$\ln[k] = -\frac{Ea}{RT} + \ln[A] \tag{2.15}$$

2.2.3 Transesterification Reaction

WCO was obtained from a local restaurant in Johannesburg. Methanol (purity: 99%), iso-propanol (purity: 99%) and sodium silicate (the catalyst) used in the transesterification of WCO oil to biodiesel were purchased from Sigma Aldrich (Pty) SA and were used without any further purification. The procedure and some of the reaction parameters used for the transesterification reaction were obtained and modified from elsewhere [14–16]. The reaction was carried out in an FBR whose process flow is depicted in Figure 2.2. The catalyst (sodium silicate) was calcined at 200°C for two hours to remove moisture and other organic matters like oil. Then the FBR was packed with 4.4 g of the catalyst diluted with 51 g inert glass beads. The FBR was then immersed in a water bath which was set at different reaction temperatures between 30°C and 60°C for the different experimental runs. Alcohol-to-oil molar ratio of 9:1 was used in the reaction. To preheat the mixture and allow uniform heat distribution, the reactor was held by a clamp on a retort stand and immersed in a water bath. Peristaltic pump was used to pump the pre-heated feed mixture into the FBR at flow rate of 2.2 L/min. The effluent of the reactor was recycled into the reactor to enhance conversion. Four experimental runs at different temperatures were carried out for 90 minutes for each of the temperature changes, and samples were taken for analysis at an interval of 30 minutes. A pre-calibrated gas chromatograph

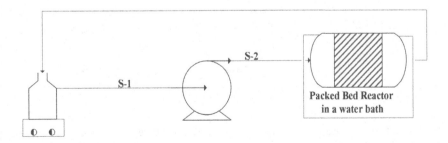

FIGURE 2.2 Schematic diagram for transesterification reaction experimental setup.

(GC) equipped with a flame ionization detector (FID) and helium (He) as the carrier gas was used to analyse the samples withdrawn from the reactor. The biodiesel yield and WCO conversion were obtained using Eqs. (2.16) and (2.17), respectively:

$$\% \ Yield = \frac{C_{FAME \ (final)}}{C_{TG \ (initial)}} \times 100 \tag{2.16}$$

$$\% \ Conversion = \frac{C_{TG \ (initial)} - C_{TG \ (final)}}{C_{TG \ (initial)}} \times 100 \tag{2.17}$$

2.3 RESULTS AND DISCUSSION

2.3.1 CONCENTRATION OF TRIGLYCERIDE AND FAME WITH TIME

The conversion of the TG in the reactor as a function of time and temperature obtained experimentally is shown in Figure 2.3. As can be seen from the figure, the concentration of WCO (TG) reduced with time in the reactor, indicating the consumption of the reactant and conversion into products in the reactor. It could be seen that the reaction rate is enhanced at higher temperature. This is expected because increase in temperature enhances rate of reaction.

Equation (2.11) was used to estimate the reaction rate constant at different temperatures used in the developed model by plotting $\ln[C_{TG(t)}]$ against reaction time (t). Reaction constants at different temperatures were estimated from the slope of the plot for each temperature. Figure 2.4 depicts the plot of $\ln[C_{TG(t)}]$ versus reaction time, and Figure 2.5 shows a plot of $\ln[k]$ against the inverse of temperature from which the activation energy was estimated. The activation energy was estimated by linearizing the Arrhenius equation. $\ln[k]$ was calculated from the estimated k values and plotted against the inverse temperatures 30°C, 50°C and 60°C.

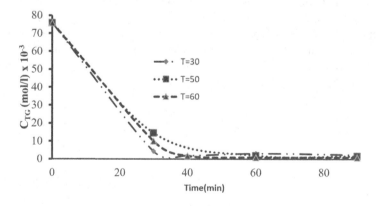

FIGURE 2.3 Concentration of triglyceride versus reaction time. *Experimental conditions:* Methanol: Oil ratio: 9:1; Catalyst weight: 4.4 g; Reaction temperature: 90 minutes; Reaction temperature: 30°C, 50°C and 60°C.

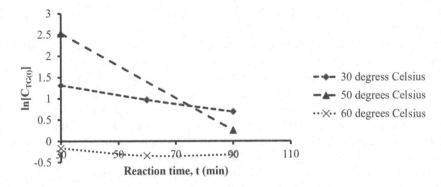

FIGURE 2.4   The plot of ln[CTG] vs. reaction time. *Experimental conditions:* Methanol: Oil ratio: 9:1; Catalyst weight: 4.4 g; Reaction temperature: 90 minutes; Reaction temperature: 30°C, 50°C and 60°C.

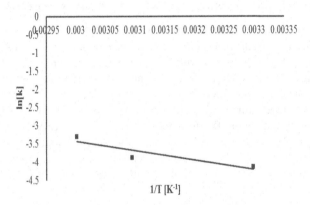

FIGURE 2.5   The plot of ln[k] vs the inverse of temperature to estimate activation energy for the forward reaction.

The estimated reaction rate constants at the respective reaction temperature as compared to literature are presented in Table 2.2. It can be observed that these values are very low when compared to the values obtained from literature [17]. This deviation can be attributed to the fact that rate constants obtained from literature are for homogenous catalysts, while the catalyst used in this study is heterogeneous.

TABLE 2.2

Estimated Kinetic Parameters Compared to Literature

Parameters	Temperature (°C)			Activation Energy (kJ/mol)	Reference.
	30	50	60		
Reaction rate constant (s^{-1})	0.011	0.038	0.003	21.46	This study
Reaction rate constant (s^{-1})	0.242	0.215	0.242	19.12–34.92	[18, 19]

According to Le Chatelier's principle, an increase in reaction temperature will increase the kinetic energy of reacting molecules, thus agitating molecular collisions in the reactor. This leads to the assumption that increase in temperature will increase the rate constant, further affirmed by the Arrhenius equation. The rate constant estimated for a temperature of 60°C deviated a bit from the stated principle, and there is no clear reason for this deviation, but it can be attributed to experimental error. The activation energy was estimated to be 21.46 kJ/mol, and this indicates the energy per mole of reactant needed to facilitate the reaction. Aransiola et al [18, 19] reported an activation energy range of 19.12–34.92 kJ/mol in a kinetic study of homogenously catalysed transesterification reaction. The activation energy estimated in this study falls within this range provided by literature. Therefore, the activation energy can be assumed to be acceptable according to literature. The estimated kinetic parameters were employed in the solution of the developed model.

2.3.2 CONCENTRATION PROFILE AND SENSITIVITY ANALYSIS FROM THE MODEL

Figure 2.6 shows the concentration profile as a function of the length of the reactor for the reactants and products as obtained from the solution of the reactor model. As could be seen in Figure 2.6, concentration of methanol (A) and TG (B) decreased along the length of the reactor. This is because they are being consumed as the transesterification reaction proceeds in the reactor. The longer the reactor length, the higher the contact time between the reactants and the catalyst at a fixed feed flow rate, thereby enhancing the formation of the products (FAME [C] and glycerol [D]). Figure 2.6 shows a comparison between FAME production obtained from experimental results and FAME production reported in literature. Qualitatively, the observation from the model solution agrees with literature [9, 11].

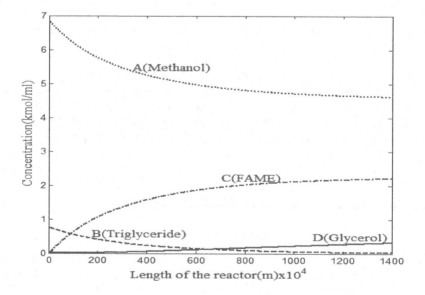

FIGURE 2.6 Concentration profile in the FBR for methanol, triglyceride, FAME and glycerol.

At a reaction temperature of 50°C, sensitivity analysis of the model was carried out to understand the behaviour of the reactor to perturbation. In the sensitivity analysis, a small (±25%) change in the inlet WCO, ±25% in the reactor radius and ±25% change in the reactor length were considered. The idea was to gain an insight into the behaviour/response of the reaction/reactor to the perturbations. During this investigation, other variables were considered constant. Changes in conversion and biodiesel yield were observed and compared to the original model output. Results of the sensitivity analysis are presented in Figures 2.7–2.9.

Figure 2.7 illustrates how sensitive the reaction performance is with respect to ±25% change in the inlet concentration of the WCO. According to the model, at a +25% change in the WCO, the WCO conversion was reduced by about 4% when compared to the reference concentration. In addition, the WCO conversion increased from 97.32% to 98.50% when the inlet concentration of WCO was reduced by 25%.

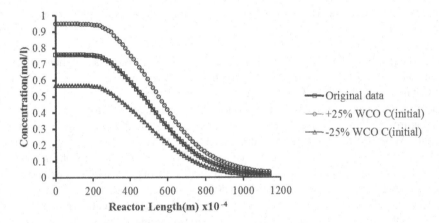

FIGURE 2.7   Concentration profile inside the reactor with respect to ±25% change in inlet WCO (feed flow rate).

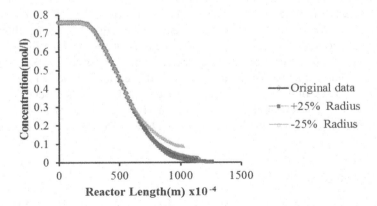

FIGURE 2.8   Concentration profile inside the reactor with respect to ±25% change in reactor radius.

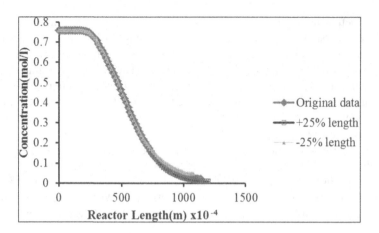

FIGURE 2.9 Concentration profile inside the reactor with respect to ±25% change in reactor length.

This means that reducing the initial concentration of WCO favours a higher conversion. This could be attributed to the fact that reducing the initial concentration of WCO increases the alcohol: oil ratio. In the presence of excess alcohol, it is easier for the reaction to approach completion and thus give a high conversion.

Figure 2.8 shows the results of the sensitivity analysis at ±25% change in reactor length. When the reactor length was decreased by 25%, the conversion was reduced to 93.9%. But when the reactor length was increased by 25%, the conversion increased to 98.8%. This is expected, as the conversion in an FBR increases with an increase in reactor length [9, 11]. This observation could be attributed to the fact that as the reactant moves along the length of the bed at constant flow rate and enhanced residence time. At the same time the conversion is enhanced with contact time where the catalyst propagates the reaction towards its completion. From this we gathered that the conversion of triglyceride to biodiesel can be achieved by increasing the reactor length.

Figure 2.9 illustrates the results of the sensitivity analysis at ±25% change in reactor radius. From Figure 2.9, reducing the reactor radius by 25% substantially reduced the conversion to 88.5% while increasing the radius by 25% resulted in a higher WCO conversion of 99.57%. This can be attributed to the fact that increasing the radius of the reactor results in an increase in the cross-sectional area of the reactor and thus an increase in reactor volume. If the reactor volume increases, there will be an increase in the contact time between the catalyst and the reactants, resulting in enhanced conversion. Results obtained during the sensitivity analysis agree with literature [9, 11], and this information could be instrumental to designing and controlling the process.

2.4 CONCLUSION

Mathematical modelling and sensitivity analysis of an FBR during the transesterification of WCO over solid sodium silicate catalysts were investigated in this chapter. The kinetic parameters employed in the modelling were obtained from experimentation as described in this chapter. Quantitatively in term of WCO conversion, a wide difference between the model result and experimental result (60% for experiment as compared to 97.3% from the model) was observed, indicating that the developed model could not accurately describe the behaviour of the reactor due to some assumptions employed during the model development. However, the trends observed for the model results and the experimental results are similar. The sensitivity analysis reveals that changes in the initial concentration of the WCO, the reactor length and the reactor radius have pronounced effects on the transesterification reaction, and the observation agrees with literature.

REFERENCES

1. 2013 Energy Statistics Yearbook. Total Energy Consumption. New York: United Nations, 2015, 46.
2. Atadashi, I.M., Aroua, M.K., Abdul, A.R. The effects of catalysts in biodiesel production: A review. *Journal of Industrial and Engineering Chemistry*, 2013, *19*, 14–26.
3. Cox, T. *BIOS Scientific Publishers*, 2nd Ed. 2004. New York: Courier Dover Publications.
4. Likozar, B., Cevec, J. Transesterification of canola, palm, peanut, soybean and sunflower oil with methanol, ethanol, isopropanol, butanol and tert-butanol to biodiesel: Modelling of chemical equilibrium, reaction kinetics and mass transfer based on fatty acid composition. *Applied Energy*, 2014, *123*, 108–120.
5. Haigh, K.F., Vladisavljević, G.T., Reynolds, J.C., Nagy, Z., Saha, B. Kinetics of the pretreatment of used cooking oil using Novozyme 435 for biodiesel production. *Chemical Engineering Research and Design*, 2014, *92*, 713–719, https://doi.org/10.1016/j.cherd.2014.01.006
6. Brásio, A.S., Romanenko, A., Leal, J., Santos, L.O., Fernandes, N.C. Nonlinear model predictive control of biodiesel production via trans-esterification of used vegetable oils. *Journal of Process Control*, 2013, *23*, 1471–1479, https://doi.org/10.1016/j.jprocont.2013.09.023
7. Khanna, R., Seinfeld, J.H. Mathematical modelling of packed bed reactors. *Advances in Chemical Engineering*, 1987, *13*, 113–189.
8. Lordanidis, A.A. *Mathematical Modelling of Catalytic Fixed Bed Reactors*, 2002. Enschede: Twente University Press.
9. Daramola, M.O., Burger, A.J., Giroir-Fendler, A. Modelling and sensitivity analysis of a nanocomposite MFI-alumina based extractor-type zeolite catalytic membrane reactor for m-Xylene isomerization over Pt-HZSM-5 catalyst. *Chemical Engineering Journal*, 2011, *171*, 618–627.
10. Perry, R.H., Green, D.W. *Perry's Chemical Engineer's Handbook*, 7th Ed. 1997. New York: McGraw-Hill.
11. Fogler, H.S. *Elements of Chemical Reaction Engineering*, 4th Ed. 2006. New Jersey: John Wiley & Sons, Prentice-Hall Inc.
12. Saltelli, A., Chan, K., Scott, E.M. *Sensitivity Analysis*, 2000. Chicester: Wiley.
13. Njomo, D., Daguenet, M. Sensitivity analysis of thermal performances of flat plate solar air heaters. *Heat and Mass Transfer*, 2006, *42*, 1065–1081.
14. Wong, Y.C., Devi, S. Biodiesel production from used cooking oil. *Oriental Journal of Chemistry*, 2014, *30*, 521–528, https:/doi.org/10.13005/ojc/300216
15. Padhi, S.K., Sing, R.K. Non-edible oils as the potential source for the production of biodiesel in India: A review. *Journal of Chemical and Pharmaceutical Research*, 2011, *3*, 39–49.

16. Hindryawati, N., Maniam, G.P., Karim, M., Chong, K.F. Trans-esterification of used cooking oil over alkali metal (Li, Na, K) supported rice husk silica as potential solid base catalyst. *Engineering Science and Technology, an International Journal*, 2014, *14*, 95–109, https://doi.org/10.1016/j.jestch.2014.04.002

17. Noureddini, H., Zhu, D. Kinetics of transesterification of soybean oil. *Journal of the American Oil Chemists' Society*, 1997, *74*, 1457–1463.

18. Aransiola, E.F., Ojumu, T.V., Oyekola, O.O., Madzimbamuto, T.F. Ikhu-Omoregbe D.I. A review of current technology for biodiesel production: State of the art. *Biomass and Bioenergy*, 2013, *61*, 276–297.

19. Aransiola, E.F., Daramola, M.O., Ojumu, V.T., Layokun, S.K., Solomon, B.O. Homogeneously catalyzed transesterification of Nigerian *Jatropha curcas* oils to Biodiesel: A kinetic study. *Modern Research in Catalysis (MRC)*, 2013, *2*, 83–89.

3 Phenomenological-Based Semi-Physical Models
Formalizing the Tool

Hernan D. Alvarez

3.1 INTRODUCTION

To the author's knowledge, no complete studies have focused on characterizing the phenomenological-based semi-physical models (PBSMs), in spite of their use in chemical and biochemical processes. The PBSM family, one of the three possible model families, does not have a formal recognized characterization and a rigorous modeling methodology within all those who practice the process engineering. Perhaps for this reason, there is no widespread acceptance of its existence as a specific class of models. Therefore, most of the published models based on phenomenological knowledge are complex to understand and some parts of their deduction are obscure and non-evident to the reader. Obviously, there exist some published works with good intention containing proposed methodologies to obtain PBSMs.

In spite of being an excessively treated topic in the literature, to write about modeling and particularly on PBSM, is not the rediscovery of the wheel. The PBSM had been used in engineering at least from the 1960s, but the discussion about model characteristics and formalization has not been considered with enough detail and completeness. Present work tries to contribute to the beginning of a wider discussion about PBSM and their formalization, until now an uncompleted task in process engineering.

This work presents the concept of phenomenological-based semi-physical models (PBSM) and exposes a methodology to obtain a PBSM for a given process. In addition, the role of models in process engineering, considering the model as the key piece for knowledge generation, storing, analysis, and other model-based engineering tasks, is presented in brief. To state the models differences, a model classification is presented, but emphasizing on the PBSM. Outstanding characteristics of this family of models are discussed, illustrating the advantages of having a PBSM to provide additional process information. A procedure to obtain PBSM is explained in detail and applied to a known benchmark, the continuous stirred tank reactor (CSTR). Finally, two critical steps of proposed methodology are presented for various published models to illustrate the wide possibilities of PBSM supporting model-based tasks in process and biological engineering.

DOI: 10.1201/9781003435181-3

3.2 A BRIEF ABOUT MODEL USES IN PROCESS ENGINEERING

Models are representations of real objects in a useful format for a given task in accordance with user requirements. The mathematical models are the most popular format to represent real processes, but other models possibilities exist like verbal or graphical ones. When process engineering tasks are developed, the availability of a dynamic model for the process being treated is an evident advantage. If, in addition, such a model has interpretability from the phenomena taken place in the process, its usefulness in design, control, and optimization is outstanding. This desire had conducted a major stream in process engineering research from the middle of the 20th century, with pioneer books like those of Bird and co-workers [1] and Rudolph Aris [2], both widely known. However, a discussion about the best kind of possible models to be used in a particular process engineering application and their specific capabilities had been forgotten or neglected due to its apparently high theoretical charge. This sensation may be the result of the high load of mathematics put in the treatment of transport phenomena when phenomena foundations are explained [1]. However, that mathematical charge can be lightened but not forgotten, when phenomena are modeled for process engineering uses. Mentioned discussion about models really is not heavy or stressing when it is done from practical process applications at the macroscopic scale.

No any model is useful for any particular application in process engineering. For example, it is impossible to use a static model to do a dynamic analysis. On the other hand, it is evident that some activities in process engineering can be executed from a small model containing the basic dynamic behaviors of the process. But, other activities require a big model with the largest number of dynamic behaviors represented in it. The difference between using a small or a large model is the knowledge gained about the process treated during model construction or interpretation. This leads to differentiate between interpretable and poor interpretable process models. Obviously, the time-cost trade-off of model construction plays a critical role in deciding the kind of model to use. However, it is better to have the possibility of obtaining both kinds of models, the minimum and the maximum reachable models, counting with current knowledge resources. In this way, the engineer could evaluate both, short and big model construction costs.

Short models, like transfer functions in Laplace domain or auto-regression, normally come from data correlations. These models are documented extensively in classic literature [3]. On the other hand, large or detailed models proceed from phenomena analysis and are less treated or documented in published works. This lack of documentation about phenomenological-based models has produced a curtain of mystery and complexity around that models. If this curtain is removed, all the possibilities of this kind of models are available for biological and process engineering practitioners.

An extensive enumeration of model applications in process engineering is not required to state three main characteristics of an ideal model: *i*) has enough accuracy, *ii*) contains maximum interpretable parameters, and *iii*) comes from an easy to follow modeling procedure. In addition, as valuable secondary characteristic of a model, it should provide enough information beyond typical prediction results. Some extra information obtained from the process model without the need to solve it will be useful for improving any model-based application, design, optimization, or control. All these characteristics are present in PBSMs, the kind of model discussed in this chapter.

3.3 MODEL FAMILIES AND THEIR CHARACTERISTICS

Models in process engineering are useful mathematical representations of interesting behaviors of the process. In that sense, any mathematical function can act as a process model. The origin of that function and its capabilities to represent available knowledge about the modeled process provide particular properties to the model. Despite apparent wide model possibilities, only three families of models exist when the origin of the mathematical structure of the model is taken as classification criterion: *i*) phenomenological or physical or white box models, *ii*) empirical or black box models, and *iii*) semi-physical or gray box models. In the following, each family is discussed for establishing differences among them. A common characteristic of all models is the use of a prediction error (PE) as a numerical guide to make the model parameter adjusting, looking to improve the model accuracy. The PE is defined as the difference between experimental data value and the value of model predictions for the same conditions.

3.3.1 PHENOMENOLOGICAL MODELS

Phenomenological or physical models use theories, laws, or principles as the only sources for proposing their mathematical structure [1]. In that sense, this kind of models contains all the required knowledge about the modeled behaviors of the process. Thus, these models are known as physical or white box models, meaning they are not contaminated with black points of uncertainty about real phenomena taken place in the process. Any mathematical expression used for evaluating parameters in a phenomenological or physical model must be obtained from mentioned sources, theories, laws, or principles [4]. This fact reports the main characteristics of phenomenological models: *i*) totally explanatory models without assumptions about reality of process phenomena and *ii*) difficult to find, mainly due to mentioned requirement about the origin of all model equations (including equations for model parameters).

An outstanding characteristic of phenomenological models is their inherent interpretability of model parameters. Due to the origin of mathematical model structure and parameters equations, the model parameters always have interpretation in the context of the modeled process. Theories, laws, and principles used to construct such a kind of models provide inherent parameters interpretability from a general knowledge context or from a particular context, restricted to the family of process being modeled. This characteristic is partially inherited by any phenomenological-based model, i.e., any model using a basic model structure deduced from phenomenological knowledge. Finally, it should be mentioned that phenomenological models are so rare in the literature due to its distinctive feature of being only formed from general knowledge about the process: theories, laws, and principles. Therefore, this family is little used in process engineering, except for exemplifying some basic phenomena [1].

Frequently found works in the literature abuse the term phenomenological using it to denote models not totally white box; i.e., they contain parameters evaluated through numerical correlations, which are not obtained from theories, laws, and principles. Therefore, these correlations do not have interpretability from the phenomena occurring in the process. In that sense, mentioned models are not pure phenomenological models. They should be called phenomenological-based models, a different family of models, as it will be explain below.

3.3.2 EMPIRICAL MODELS

This kind of models is totally opposite to the previous one because here a total absence of knowledge about the phenomena taken place in the process is assumed. This fact gives to this kind of models their most popular name: black box model, meaning they do not have an interpretable knowledge about the process phenomena being modeled. With this license, when an empirical model is being constructed, any mathematical function can be used as model structure. The only requirement for proposed mathematical function is its capability of reducing the model PE by adjusting structure or parameters of the model [3].

Finding black box or empirical models in the literature is a very easy task. In the same way, they are very easy to obtain owing to their advertising feature of "plug and play" models. In that sense, there are online tools for obtaining an empirical model giving a dataset. Thus, empirical models are the majority of models reported when a given field of the science is in exploratory and correlation phase of development. But, when the field is in descriptive or explanatory phase, empirical models are totally absent, being replaced by model of the next family, semi-physical models. The main characteristics of empirical models include the following: *i*) these models are quick and easy; i.e., these are quick to obtain and easy to use, and *ii*) these models do not have parameters interpretability, which strongly limits their uses in research. To close this brief discussion, it should be highlighted that empirical models can be called purely predictive models. Under the evident fact that any model has prediction capability, the "purely" adjective indicates empirical models only have this capability, opposite to phenomenological models which have another capabilities in addition to the prediction functionality.

3.3.3 SEMI-PHYSICAL MODELS

This family profits the outstanding capabilities of the two previous families mixing the properties of white and black box model, hence their name of gray box models. Two subfamilies appear in accordance with model structure origin: *i*) empirical-based models or *ii*) phenomenological-based models. A model of the first subfamily takes its basic structure from an empirical model (black box), while a phenomenological-based model takes its basic structure from a phenomenological model (white box). The first subfamily is used very little due to the rigid mathematical structure of any model of this kind. That rigidity is inherited from the empirical model that originates the basic structure. Therefore, it is possible to say that empirical-based semi-physical models are theoretical object without practical uses. On the other hand, the

PBSMs are gaining recognition among the process engineers due to the publication of methodologies to obtain them and examples of PBSM used in chemical and biological processes. This kind of models takes the best of phenomenological models (basic structure from theories, laws, and principles), and the best of empirical models (quick and easy sub-models as an option for calculating model parameters). PBSM are the most useful sub-family of gray box models due to that successful combination, as it is shown in the rest of this chapter.

3.4 PHENOMENOLOGICAL-BASED SEMI-PHYSICAL MODELS (PBSM)

As a brief definition, it can be said that a PBSM is a subtype of gray box models with model structure taken from first principles, but including at least one model parameter calculated with an empirical expression or sub-model. In that sense, a PBSM inherits all desirable properties of phenomenological models, with the advantage of being optionally able to find model parameters through empirical equations. The mathematical structure of a PBSM is found by applying the conservation principle, as it is done for phenomenological models (white box). This structure, called *model basic structure* consists of those balance equations (BEs) selected among all obtained balances looking for the simplest balances with enough information to fulfill the model objective. Parameters in model basic structure are indicated by symbols. The mathematical expressions for those parameters must not be replaced into the model basic structure looking for maintaining the inherent interpretability of the PBSM basic structure. This set of parameters are called *structural parameters* to distinguish them from the rest of parameters to appear when mathematical functions are used to calculate those structural parameters. These other parameters are called *functional parameters* due to its origin from mathematical functions different to balances.

In general, model parameters are represented either by mathematical formulations or by numerical values, fixed by the modeler or identified by minimizing model PE [3]. The set of equations for structural or functional parameters form the *model extended structure*. Unlike model basic structure, which is unique for the kind of process to which the modeled process belongs, the model extended structure is particular and strongly depends on modeler knowledge. Mentioned characteristic of model basic structure is valuable in extreme to provide a bank of PBSMs for unit operations and other single processes as electric machines [5] or mechanical devices [6]. The model basic structure plus the model extended structure form the *(total) model structure.*

Considering a PBSM as a mathematical system, its main parts are the terms, representing an input or output flow, or an internal change on time of mass, energy, momentum, or any other interesting property. These quantities are evaluated into a given volume or as the mass, energy, momentum, or interesting property flow passing through one of the walls of that volume. A term can be formed by variables, parameters, and constants. A model variable is characterized because its value cannot be determined a priori, requiring model solving to know its value. Contrary, the parameters will never be the result of model solution. They need to be specified by the modeler previous to model solution. It is recommended to call the parameters in accordance with their level in model, structural or functional. It is evident that different to parameters, the model variables are ever on structural level. The constants of the model are numerical values representing universal quantities or process properties fixed at given values by the modeler. A particular characteristic of PBSM is that, from the point of view of terms, the model is linear because each balance equation is a sum of terms, each one with positive or negative sign. However, from the point of view of variables and parameters coexisting into model terms, nonlinear characteristics are evident in most of the PBSM. To know this difference, let's treat the terms as linear parts and the variables and parameters as nonlinear ones. The bulk of identification procedures profit this difference to improve the model parameters identification [3].

3.5   PBSM OUTSTANDING CHARACTERISTICS

Not all mathematical structures used as models offer the same representation capabilities when tasks beyond pure prediction are intended. From PBSMs, four main characteristics are recognized: *i*) uniqueness of model basic structure, *ii*) modularity of model extended structure, *iii*) option of combining different levels of detail, and *iv*) model parameters interpretability. A brief discussion about these characteristics will help to identify very interesting capabilities of PBSM when compared to empirical ones. No mention is required to phenomenological- and empirical-based semi-physical models. The phenomenological models inherently have all mentioned properties and empirical-based semi-physical models are rarely used in engineering application, as it was previously said.

Uniqueness of basic model structure in PBSM is evident because the basic model structure comes from applying the conservation principle, which is a major phenomenological description. Such uniqueness means that a family of processes has the same basic model structure. For example, every heat exchanger has the same basic model structure. The difference between two heat exchangers resides in model parameters, just the extended model structure. The equations for model parameter are different for each heat exchanger due to the effect of constructive particularities over heat transfer coefficients. The same can be applied to mass transfer pieces, chemical or biological reactors, electrical machines, mechanical assemblies, etc.

The modularity of model extended structure indicates the possibility of replacing an identified value or a given function for evaluating a model parameter by a new parameter evaluation function. In this way, the model extended structure is increased without modifications of the previous part of that structure. The new function is plugged into the model position corresponding to the previous identified value or evaluation function. In that way, the model can grow as new information or knowledge about the process is gained, without dramatical changes in model structure, as it happens with empirical models when new information is available [3]. For example, the replacement of a viscosity value found by identification from data by an explicit expression to calculate that viscosity is direct [7].

To combine different levels of detail in a PBSM is a property directly linked to the previously mentioned modularity. The option of combining information in macroscopic level with information obtained in lower levels as mesoscopic or microscopic is direct. Modularity allows to replace a fixed value of a parameter of upper level for a mathematical expression based on new parameter(s) of a lower level, allowing to calculate the upper parameter value. For example, to evaluate from bubble properties at low size level the mass transfer coefficient to use in upper macroscopic level [8]. Obviously, that combination is fruitful when the lower levels provide knowledge about the phenomena instead of only giving information about the parameter value. This fact is extended in the next property, the interpretability. Level of detail should not be confused with level of specification. The level of detail refers to the size of a partition over the process, while the level of specification refers to the quantity of knowledge or information contained in an expression used to calculate a parameter.

Finally, the interpretability of model parameters refers to the capability to give a meaning to a given model parameter. Phenomenological models have the highest interpretability because all their parameters are interpretable because they come from theories, laws, and principles. Obviously, a PBSM inherits the interpretability for all structural parameters but the interpretability of functional parameters is not guaranteed. However, contrary to an empirical model, a PBSM has the outstanding capability of increasing parameters interpretability provided the parameters evaluation functions being selected with phenomenological origin or inspiration [9].

3.6 A PROCEDURE TO OBTAIN A PBSM

By integrating known procedures to obtain phenomenological-based models, it was possible to propose a procedure to construct PBSMs. The procedure was developed by interacting with model users and reviewing published procedures as in [10] and [11], citing only 2 among 21 reviewed works (see [12]). The idea was to simplify some steps in existing procedures, looking for a direct sequence easily applicable by junior engineers without mandatory requirement of deep knowledge about the process. The procedure clarifies some obscures steps of published procedures by modifying existing steps or by introducing new ones. Mentioned procedure had been successfully used to obtain process models in several works [13–15]. The next 10 steps of the modeling procedure are presented with an explanation of each one, followed by step application to a simple but interesting industrial process, a CSTR. In spite of the wide treatment available in published works about CSTR, in these works the CSTR model deduction is treated in brief or is presented in a reduced way or without following a formal modeling method. Here, a complete and detailed CSTR model deduction is presented, looking for a thorough understanding of PBSM procedure and CSTR phenomena interactions. Despite being a simple process, a CSTR contains enough complexity to illustrate the usefulness of proposed modeling procedure. In addition, final section of this chapter illustrates several model hypothesis used in PBSMs looking for exemplifying that critical step in the proposed methodology.

3.6.1 STEP 1: PROCESS DESCRIPTION AND MODEL OBJECTIVE DECLARATION

This step indicates the formulation of the question to be answered by the model, known as model objective, and to write a verbal description and sketch a process flow diagram (PFD) that complement each other. This step is important because

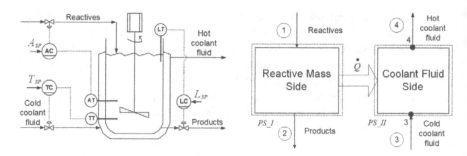

FIGURE 3.1 *P&ID* of modeled CSTR on left side and block diagram for process systems taken for PBSM on right side.

a poor understanding of modeled process leads to an erroneous model. Mentioned description must be detailed enough but avoid irrelevant aspects of real process. It is expected that the verbal description and the PFD will suffer refinements during model construction. It is important to highlight that the model objective is at least one question to be solved by the model, not to be solved by a model user profiting the model responses or solution.

As example, the process modeled here is a reactor conducting an exothermic reaction $A \rightarrow B$ in continuous fashion into an agitated tank, so this process is called CSTR. The reactor tank is a vertical cylinder, i.e., constant cross area. Inlet flow of reactive solution and outlet flow of product, with remaining reactive solution, are called streams 1 and 2, respectively. Surrounding the reactor tank a jacket for circulating coolant liquid is assembled. Input and output flows of coolant are tagged streams 3 and 4, respectively. Temperature and reagent A concentration in reactor feed stream are known. Pressure and temperature of coolant fluid at jacket inlet are known too. Left side of Figure 3.1 illustrates the piping and instrumentation diagram (*P&ID*) of the process. The (*P&ID*) uses ISA norm implying that concentration controller appears as an Analysis Controller (*AC*), although in the rest of the text is named *C* for indicating concentration of a given stream. Right side of the figure will be discussed in next step. Process conditions and extensive description are presented in [16]. The model objective is to answer the next questions: *i*) how does the concentration of reagent A in reactor outlet stream change over time? *ii*) How does the temperature of reactive mass change over time? *iii*) How does the temperature of outlet coolant stream change over time? *iv*) How does the outlet pressure of coolant stream change over time?

3.6.2 STEP 2: MODELING HYPOTHESIS AND LEVEL OF DETAIL

Developing this step implies to state how the operation is and what phenomena take place in the actual process or an analogy to it. An analogy is a knowledge-based guesswork, i.e., to use the imagination to describe what is unknown. Obviously, the imagination is based on what is known, which means the imagination is not a totally free flight. In this way, the unknown part described using the imagination gains similarity with a known object, although there is some uncertainty about the credibility of such similarities. That credibility begins to be certified by the results of the analogy when the measured behavior of an unknown part can be predicted using what the imagination build as the phenomena taken place. Following this line, the modeling hypothesis is a short description of the real or the supposed way in which an enough number of known phenomena are interacting to make true the current process being described. When the real phenomena of the modeled process are not fully known, it is possible to take a very well-known process as analogous to the currently modeled process. Examples of this situations are presented in the last section of this chapter.

To construct the modeling hypothesis, it is possible to relax one or more known phenomena conditions in order to make it useful for describing the interesting part of the described real process. To this aim, a set of assumptions justifies that assembly of known phenomena, in the real process or the analogous one. An assumption is a single sentence stating a condition or value for an object, a simple operation, a variable, or a parameter of modeled process. Due to the permanent review of modeling hypothesis during model development, some assumptions could change. Therefore, assumptions consistency must be checked each time a new assumption is included or one existing is modified to avoid impossible situations deduced from current assumptions set.

On the other hand, to construct a model, a partition over the real process must be done to appreciate process parts and their interactions. In the case of PBSM, these parts are called process systems, as it is discussed in the next step. Obviously, a level of detail for the model according to model objective or purpose should be stated too. The level of detail is the position in a scale of sizes of useful engineering possible parts when the current partition is taken. That scale begins at gross detail, which indicates the real object as a whole differentiated from the rest of the world, and finishes at the minimum physical size (atomic, for example) possible to use for the modeled process. Several levels of detail exist in a PBSM but only the level at which the bigger parts, called process systems, are determined gives the level of detail of the model. Level of detail should not be confused with lumped of distributed parameters (see Step 6) or with specification level (quantity of available knowledge) previously discussed. As can be seen, steps 1 and 2 are a loop with permanent improvement during model development.

Applying all these aspects of Step 2 to our example, the CSTR, the phenomena taken place in this process are known from textbooks; therefore, no analogy for process operation is required. In addition, the modeling hypothesis is simple: molecules of reagent A dissolved in solvent S begin their exothermic transformation into molecules of product B due to energy and agitation conditions of reactive mass into the tank. The heat transfer takes place from the reactive mass into tank to the coolant fluid flowing through the jacket. Next assumptions complete the hypothesis:

- Perfect agitation condition is guaranteed in tank and jacket.
- Heat transfer between reactive mass and fluid in the jacket is due to forced convection mechanism.
- Heat exchange between jacket and surroundings is negligible.

- Kinetics of reaction $A \rightarrow B$ can be represented by Arrhenius law.
- Solvent S is inert regarding reaction taken place.
- The loss of mass from reactive mass by evaporation are negligible.
- Density changes of coolant fluid due to temperature changes are negligible.
- Heat capacities of reactive mixture into reactor are constant.
- Jacket is always full of coolant fluid.
- The controller used in the process uses the PID algorithm.

The level of detail to answer the model question is macroscopic because the model should give time trajectories of two variables of the reactive mass as a whole, concentration and temperature. In addition, under perfect agitation assumption, no spatial distribution of parameter exists, both in reactive mass and coolant flowing through the jacket.

3.6.3   STEP 3: DEFINITION OF PROCESS SYSTEMS (PS)

At this step, it is required to practice a partition over the process. This partition consists of defining as many process systems (PS) on the process to be modelled as required by the level of detail. A PS is defined as the abstraction of a part of the process as a system, i.e., over that part all available tools of systems theory and system engineering can be applied. The origin of the name *Process System* is a process part viewed as a system, as it was stated in [11]. Next, some indications to PS determination are given. *i*) Look for physical walls into the process considering that the two volumes separated by a wall could be considered two PS. For example, in a shell and tube heat exchanger, the fluid into tubes is a PS and the fluid in the shell is another PS. *ii*) Detect distinguishable phases of substance into the same volume and assign a PS to each phase no matter one of them is disperse and the other is the continuous phase. For example, in an aerated tank, the bubbles of air are grouped in a PS and the liquid is another PS [15]. *iii*) Detect any mass characteristic marking spatial differences and separate each portion of mass with similar characteristics as a PS. For example, in a bioprocess where cells could be in the vegetative or the sporulation stages, one PS is declared for all cells in vegetative stage and another PS is declared for those cells in sporulation stage [17]. *iv*) Use arbitrary limits when obtained PS are useful for model development. For example, when two metal sheets are being welded, the thermally affected portion of the sheets is declared as a PS and the not affected part of the sheets is the another PS [14]. To finish this step, a block diagram relating all PS must be plotted. Thin arrows are suggested to indicate mass flows and thick arrows for energy flows.

In the CSTR case study, two process systems are taken, as it is illustrated by a block diagram on the right side of Figure 3.1. In this diagram, the thick arrow shows the flow of heat between reactive mass and coolant fluid. The Process System I (PS_I) is the reactive mass permanently agitated into the reactor. The Process Systems II (PS_{II}) is the coolant fluid flowing through the jacket. The volume of PS_I is the mass of reagents, solvent, and products contained at a given time into the tank. That volume is variable, contrary to the volume of PS_{II}, which is constant because jacket is always full of coolant fluid, in accordance with previously stated assumptions.

3.6.4 STEP 4: APPLICATION OF CONSERVATION PRINCIPLE

This step implies to apply the principle of conservation on each defined PS to obtain a base of equations from which the model basic structure will be extracted. It is recommended to take at least the next balances over each PS: *i*) total mass, *ii*) *n*-components mass, *iii*) thermal energy, and *iv*) mechanical energy or momentum. Any other balance of interesting properties can be included. These group of equations form the BEs set. This is not a minimal set because some of those equations give the same information. Therefore, as it will be stated in next step, some of the equation could be discarded during model basic structure determination.

For the CSTR being modeled, the model basic structure is obtained after applying total mass balance, only one component mass balance for the reagent *A*, and the thermal energy balance. The balances are developed with various intermediate expressions possibly useful for model basic structure to be determined in next step. Therefore, none of them are discarded in this step, looking for having a wide bank of optional expressions. In the following, each process system is treated in one section.

3.6.4.1 Process System I (*PS$_I$*)

The balances applied to reactive mass into the reactor tank are as follows, using molar units to easily account for reaction effects. To simplify the notation in the following, the time dependence of all terms is not indicated. For example, instead of writing the total moles in time as $N(t)$, only N will appear.

Total Mass Balance. Applying a balance over total mass in *PS*:

$$\frac{dN_I}{dt} = \dot{n}_1 - \dot{n}_2 + r \sum_i \sigma_i \tag{3.1}$$

with N_I the total moles in process system *I*, i.e., the sum of moles of reagent *A*, product *B*, and solvent *S* into the tank at each moment, given in kmol. $\dot{n}_1$ and $\dot{n}_2$ are the inlet and outlet molar flows from reactor tank, respectively, given in $\frac{kmol_f}{s}$, with f the stream number. r is the reaction rate in $\frac{kmol}{s}$ and σ_i is the dimensionless stoichiometric coefficient of compound *i* taken from the balanced chemical reaction expression. σ_i is signed, using + for reaction products and − for reagents.

Equation 3.1 can be used in its current form for finding N_I value at any time. However, it is possible to express N_I in terms of density ρ_{mix} and volume V of reactive mass. This volume can be formulated in terms of tank cross area A_T and level L of liquid into the tank, resulting in

$$N_I = V\rho_{mix}\frac{1}{\mathfrak{M}_{mix}} = A_T L\rho_{mix}\frac{1}{\mathfrak{M}_{mix}}$$

where $\mathfrak{M}_{mix}$ is the molecular mass of the mix into the tank. Applying derivative operation over last expression for N, considering A_T and $\mathfrak{M}_{mix}$ as constants, produces

$$\frac{dN_I}{dt} = \frac{A_T}{\mathfrak{M}_{mix}}\left(L\frac{d\rho_{mix}}{dt} + \rho_{mix}\frac{dL}{dt}\right)$$

which, replaced in original total mass balance, gives

$$\frac{A_T}{\mathfrak{M}_{mix}}\left(L\frac{d\rho_{mix}}{dt}+\rho\frac{dL}{dt}\right)=\dot{n}_1-\dot{n}_2+r\sum_i\sigma_i$$

Considering that a total mass balance should have as main variable one related to mass accumulation and level is the accumulation variable for liquids, the differential of this equation is $\frac{dL}{dt}$, with $\frac{d\rho_{mix}}{dt}=\dot{\rho}_{mix}$ a parameter instead of differential operator. Parameter $\dot{\rho}_{mix}$ is determined from other expression and therefore moved to right side, producing the last total mass BE for PS_l:

$$\frac{dL}{dt}=\frac{\mathfrak{M}_{mix}}{\rho_{mix}A_T}\left(\dot{n}_1-\dot{n}_2+r\sum_i\sigma_i\right)-\frac{L}{\rho_{mix}}\dot{\rho}_{mix} \tag{3.2}$$

Component Mass Balance. Reagent A. The mass balance by component A in molar units is

$$\frac{dN_A}{dt}=x_{A,1}\dot{n}_1-x_{A,2}\dot{n}_2+\sigma_A r \tag{3.3}$$

where N_A is the total mass of component A in kmol. It should be highlighted that complete notation is $N_{A,I}$ for indicating that moles of A are evaluated only in PS_I. However, considering A is only PS_I, the subindex I is not used here. The symbol $x_{A,i}$ is the A concentration in molar fraction for the reactor tank streams $i=1,2$, and σ_A the stoichiometric coefficient of A in the balanced chemical reaction, recalling that, in this case, it is negative because A is the reagent.

As for the total mass balance, Eq. 3.3 could give the total number of moles of A at any time. However, it is useful to replace N_A in terms of total moles into reactor and A concentration in mole fraction of leaving stream $x_{A,2}$ under perfect agitation assumption: $N_A=x_{A,2}N_I$. Applying the derivative operator over this expression:

$$\frac{dN_A}{dt}=x_{A,2}\frac{dN_I}{dt}+N_I\frac{dx_{A,2}}{dt}$$

which is replaced in original component balance expression, produces

$$x_{A,2}\frac{dN_I}{dt}+N_I\frac{dx_{A,2}}{dt}=x_{A,1}\dot{n}_1-x_{A,2}\dot{n}_2+\sigma_A r$$

where considering that interesting variable for a component A mass balance should be the concentration of A, the change of total moles into PS_I regarding time $\dot{N}$ is a parameter, moved to the right hand. Note that $\dot{N}_I$ is evaluated using the total mass balance, Eq. 3.1. Therefore, the final component mass balance for A is

$$\frac{dx_{A,2}}{dt}=\frac{1}{N_I}\left(x_{A,1}\dot{n}_1-x_{A,2}\dot{n}_2+\sigma_A r-x_{A,2}\dot{N}_I\right) \tag{3.4}$$

Component Mass Balance. Product B. As three components form the reactive mixture, only another component mass BE is required. In the case for Product B, the equation is summarized here because the procedure is identical to that of A:

$$\frac{dx_{B,2}}{dt} = \frac{1}{N_I}\left(x_{B,1}\dot{n}_1 - x_{B,2}\dot{n}_2 + \sigma_B r - x_{B,2}\dot{N}_I\right) \tag{3.5}$$

with $x_{B,1}$ and $x_{B,2}$ the molar fraction of B at input and output streams of reactor, respectively, and σ_B the stoichiometric coefficient of B, recalling that, in this case, it is positive because B is a product.

Component Mass Balance. Solvent S. As S is not consumed or produced or evaporated, its molar fraction can be evaluated by the typical restriction of molar fraction definition, given as an assessment equation (see Step 7). This fact highlights the apparent changed role of $x_{S,2}$ because it could be considered a variable. But a detailed analysis shows that $x_{S,2}$ could be a variable if one of $x_{A,2}$ or $x_{B,2}$ is a parameter. The three unknowns cannot be variables because it is enough to know two of three variables of any process stream to express the third one as a parameter, using the algebraic definition of molar fraction.

Thermal Energy Balance. In this case, the expression for thermal energy balance considering the changes of total energy contents E of reactive mass into the reactor tank is

$$\frac{dE}{dT} = \dot{n}_1\bar{H}_{mix,1} - \dot{n}_2\bar{H}_{mix,2} - r\Delta\bar{H}_r + \dot{Q} - \dot{W} \tag{3.6}$$

with $\Delta\bar{H}_{mix,i}$ the molar enthalpy for ith stream in $\frac{kJ}{kmol}$, $\Delta\bar{H}_r$ the heat of reaction in $\frac{kJ}{kmol}$, $\dot{Q}$ the flow of heat exchanged with the jacket in $\frac{kJ}{s}$, and $\dot{W}$ the flow of mechanical energy exchanged with surroundings in $\frac{kJ}{s}$. Even though Eq. 3.6 could give the total energy E at any time, it is less used in this form. Therefore, expressing the total energy as $E = N_I\bar{C}_{V,mix}\,T$, the product of total mass in moles, the mixture heat capacity at constant volume $\bar{C}_{V,mix}$, and the reactor temperature T, and applying derivative operation under assumption of constant $\bar{C}_{V,mix}$ equal to $\bar{C}_{P,mix}$ as the mixture is a liquid:

$$\frac{dE}{dt} = \bar{C}_{P,mix}\left(N_I\frac{dT}{dt} + T\frac{dN_I}{dt}\right) \tag{3.7}$$

and replacing this expression in the thermal energy balance, we get

$$\bar{C}_{P,mix}\left(N_I\frac{dT}{dt} + T\frac{dN_I}{dt}\right) = \dot{n}_1\bar{H}_{mix,1} - \dot{n}_2\bar{H}_{mix,2} - r\Delta\bar{H}_r + \dot{Q} - \dot{W} \tag{3.8}$$

Considering that a thermal energy balance should give as variable the temperature, the change of total moles into the reactor $\frac{dN_I}{dt} = \dot{N}_I$ is a parameter to be moved to the right of the final expression. This parameter is evaluated using right side of Eq. 3.1. Due to the perfect agitation condition, the mentioned temperature is equal to the leaving stream temperature T_2. In addition, as effects of agitation produce a negligible mechanical energy flow compared to thermal energy flow when a low viscosity liquid is agitated ($\dot{W} \approx 0$), the final expression is

$$\frac{dT_2}{dt} = \frac{1}{N_I \bar{C}_{P,mix}} \left(\dot{n}_1 \bar{H}_{mix,1} - \dot{n}_2 \bar{H}_{mix,2} - r\Delta\bar{H}_r - \dot{Q} \right) - \frac{T_2}{N_I} \dot{N}_I \qquad (3.9)$$

Mechanical Energy Balance. This balance is not important for this process system (PS_I) because friction energy losses during agitation do not give information to answer the model questions. Instead, for PS_{II} this balance is important, as it is discussed below.

3.6.4.2 Process System II. PS_{II}

For the balances applied to the coolant fluid into the jacket, mass units can be used because there is no chemical reaction in that process system. The balances are found as it is illustrated below.

Total Mass Balance. For jacket this balance is trivial:

$$\dot{m}_4 = \dot{m}_3 \qquad (3.10)$$

with $\dot{m}_3$ and $\dot{m}_4$ the inlet and outlet flows of the jacket, respectively. Note that $\dot{m}_3$ is assumed as known, letting $\dot{m}_4$ has the variable of the model, in this case an algebraic variable. Equation 3.10 is the final balance expression.

Component Mass Balance. Component mass balances have not sense for PS_{II} because coolant concentration is constant during its pass through the jacket (no chemical reaction is taken place). Therefore, no mass component is required for PS_{II}.

Thermal Energy Balance. Following the same procedure used for thermal energy balance in PS_I, but without considering chemical reaction, the final expression is obtained:

$$\frac{dT_4}{dt} = \frac{1}{M_{II} \hat{C}_{P,J}} \left(\dot{m}_3 \hat{H}_{J,3} - \dot{m}_4 \hat{H}_{J,4} + \dot{Q} \right) \qquad (3.11)$$

with subindex J to indicate properties measured at jacket conditions, T_4 a representative temperature of PS_{II} due to perfect agitation assumption. M_{II} is the total mass of fluid into jacket at any time. $\hat{C}_{P,J}$ is the heat capacity at constant pressure of the fluid into the jacket.

Mechanical Energy Balance. In spite of being negligible in previous thermal balance energy, because it is lesser than thermal effects, the mechanical energy balance in PS_{II} gives information about the pressure losses affecting the flow of the coolant liquid through the jacket. While it is possible to state a dynamic mechanical energy balance, which will provide velocity values of the coolant fluid, due to the symmetric construction of the jacket, velocity is the same and easily evaluated. In addition, the interest of the model is the pressure at coolant outlet. Therefore, it is enough to use a stationary mechanical energy balance, known as Bernoulli equation, when the fluid is a liquid. This balance, taken from points 3 to 4 marked with thick black dots over PS_{II} in the right side of Figure 3.1, is as follows:

$$\frac{P_3}{\rho_J} + gz_3 + \frac{v_3^2}{2} + \eta W = \frac{P_4}{\rho_J} + gz_4 + \frac{v_4^2}{2} + h_{f3,4} \tag{3.12}$$

where subindex 3 and 4 indicate inlet and outlet point to jacket, P is the pressure in Pa, g the gravity acceleration in $\frac{m}{s^2}$, z the vertical distance of the point regarding an arbitrary zero level, v the fluid velocity in $\frac{m}{s}$, η the dimensionless efficiency of driving machine, W the energy per unit of mass supplied by the driving machine in $\frac{kJ}{kg}$, and $h_{f3,4}$ the friction losses suffered by the fluid during its pass through the jacket, i.e., from point 3 to point 4 in velocity heads or $\frac{m^2}{s^2}$.

Equation 3.12 is the general form, which should be adapted to current situation of the jacket when a driving machine does not exist ($\eta W = 0$). In addition, the value of z is considered approximately the same on both the sides of the equation and hence also the value of v on both sides, canceling out when they are on the same side of the equation. Therefore, the final expression, letting to the right P_4, the unknown value, is

$$P_4 = P_3 - \rho_J h_{f3,4} \tag{3.13}$$

considering that friction losses are calculated with an assessment equation to be determined below in Step 7.

3.6.5 Step 5: Determination of Model Basic Structure

This step indicates to select from previous found BE set those equations with information to fulfill the model objective (Step 2). They form the model basic structure. As it was seen in the CSTR example, some BE can be stationary equations but are included as model structure as they come from balances. All redundant equations are not considered into model basic structure, but they could be considered as constitutive equation for structural model parameters, despite coming from balance deduction procedure. Those new parameters and constants are functional ones but not structural because they are not in the basic model structure.

The basic model structure for the CSTR is the set of equations presented in Table 3.1. In Step 7, other equations will appear but those new equations are not part of model basic structure. They form the extended model structure. Other equations presented previously will be used as parameter equations in Step 7. It is important to remember that no more variables will appear in next steps, but it is sure that more parameters and constant will appear during the development of model extended structure.

TABLE 3.1

Balances Equations Forming the Basic Model Structure

#	Equation	Process System
1	$\dfrac{dL}{dt} = \dfrac{\mathfrak{M}_{mix}}{\rho_{mix}A_T}\left(\dot{n}_1 - \dot{n}_2 + r\sum_i \sigma_i\right) - \dfrac{L}{\rho_{mix}}\dot{\rho}_{mix}$	PS_I
2	$\dfrac{dx_{A,2}}{dt} = \dfrac{1}{N_I}\left(x_{A,1}\dot{n}_1 - x_{A,2}\dot{n}_2 + \sigma_A r - x_{A,2}\dot{N}_I\right)$	PS_I
3	$\dfrac{dx_{B,2}}{dt} = \dfrac{1}{N_I}\left(x_{B,1}\dot{n}_1 - x_{B,2}\dot{n}_2 + \sigma_B r - x_{B,2}\dot{N}_I\right)$	PS_I
4	$\dfrac{dT_2}{dt} = \dfrac{1}{N_I\bar{C}_{P,mix}}\left(\dot{n}_1\bar{H}_{mix,1} - \dot{n}_2\bar{H}_{mix,2} - r\Delta\bar{H}_r - \dot{Q}\right) - \dfrac{T_2}{N_I}\dot{N}_I$	PS_I
5	$\dot{m}_4 = \dot{m}_3$	PS_{II}
6	$\dfrac{dT_4}{dt} = \dfrac{1}{M_{II}\hat{C}_{P,J}}\left(\dot{m}_3\hat{H}_{J,3} - \dot{m}_4\hat{H}_{J,4} + \dot{Q}\right)$	PS_{II}
7	$P_4 = P_3 - \rho_J h_{f3,4}$	PS_{II}

3.6.6 STEP 6: RECOGNITION OF VARIABLES, STRUCTURAL PARAMETERS AND STRUCTURAL CONSTANTS

In this step, the variables, parameters, and constants of the model are established. Several alternative interpretations exist in published works about the concept of "variable." To avoid confusions, the term variable must be differentiated from the term unknown, commonly used to evaluate the degrees of freedom of a set of equations (see Step 8). Model variables ever are a part of unknowns of the mathematical model, but not all unknowns of the model are model variables. All parameters are unknown for the mathematical set of equations forming the model extended structure, including those parameters fixed by the modeler. Depending on the kind of question to be solved by the model (see Step 2), a lumped or a distributed parameter model appears. The first kind does not consider spatial change on parameters, contrary to distributed parameter models. For a PBSM, the variables are those solved from the basic structure of the model, provided that the values for the model structural parameters are known. Into the set of variables some are easily detected because they are in the differential terms of equations of model basic structure. These are called the state variables due to their inherent dynamic behavior. Other model variables are identified as they are at left side of the equal sign or because the variable should be solving in an algebraic equation. This second kind of variables are called mapping variables because they do not have inherent dynamic behavior but they are the result of model solution. Steps 6, 7, and 8 form a loop, but in the final report of the model these steps appear as a sequence. Therefore, it is recommended

TABLE 3.2
Variables, Structural Parameters, and Structural Constants of the Model

Variable	Total
$L, x_{A,2}, x_{B,2}, T_2, \dot{m}_4, T_4$	7
Structural Parameter	Total
$\mathfrak{M}_{mix}, \rho_{mix}, \dot{n}_1, \dot{n}_2, r, \dot{\rho}_{mix}, N_I, x_{A,1}, \dot{N}_I, x_{B,1}, \bar{C}_{P,mix}, \bar{H}_{mix,1}, \bar{H}_{mix,2}, \dot{Q}, \dot{m}_3, \hat{C}_{P,J}, \hat{H}_{J,3},$ $\hat{H}_{J,4}, P_3, h_{f-3,4}$	20
Structural Constant	Total
A_T, σ_i with $i = 1,2,3$, $\Delta \bar{H}_r$, M_{II}, ρ_J	7

to solve these three steps for each process system before advancing in the procedure. In this way, the requirement of constitutive and assessment equations and the degree of freedoms for the model extended structure will be completed as a cyclic procedure before continuing to Step 9. It is recommended to elaborate a record of this three steps to easily verify their fulfillment.

For the case study, CSTR, the parts of the model to recognize and declare in Step 6 are referred to equations of model basic structure reported in Table 3.1. The variables, structural parameters, and structural constants of the model are listed in Table 3.2.

3.6.7 STEP 7: PROPOSAL OF CONSTITUTIVE AND ASSESSMENT EQUATIONS

Step 7 of the methodology implies to propose or find from literature constitutive and assessment equations for calculating the largest number of parameters in each PS. This is the most critical step because modeler must arbitrarily choose equations to represent the model parameters. For those parameters without a constitutive or assessment equation, the modeler should identify a value from experimental data [3]. At least one of those identified parameters can act as a fine tuner of model response, provided it can be declared as a sacrifice parameter [18]. A constitutive equation approximates the response of a physical quantity to external stimuli using a law or principle. Examples of constitutive equations are Darcy's law; heat, mass, and momentum rate of transfer laws; Arrhenius law, among others. Contrary, an assessment equation is a mathematical relation to assess a parameter value, without any intention of linking in a descriptive way the calculated value to the phenomena taken place. For example, the simple assignment of the value for an equipment dimension like tank diameter ($D_T = 1.8m$) and the value for an operative condition like feed flow ($\dot{m}_{Feed} = 2.33 \frac{kg}{s}$) are assessment equations. Particularly, these two exemplified equations belong to a kind of assessment equations called trivial assessment equation.

A constitutive or assessment equation for a given parameter can produce new parameters which must be determined too. These new parameters are deeper into parameters structure layers, i.e., they are lower in the level of specification of the model. It should be remembered that specification (knowledge) level is different to detail (size) level. The first level of specification of functional parameters are those included into constitutive or assessment equations for evaluating the structural parameters. The functional parameters of second level of specification are those produced by constitutive or assessment equations evaluating level 1 functional parameters and so on for the rest of levels of specification that appear for functional parameters.

For the CSTR being modeled as case study, constitutive and assessment equations are proposed. The functional parameters produced by those equations are declared and solved after their first appearance in the model. By space limitation, only two structural parameters and the process operative conditions are treated here in detail. All model parameters with their respective constitutive or assessment equation are presented in Table 3.3. There, the end of enumeration of all expressions associated to a given structural parameter is marked by a vertical line. In addition, the type of parameter is indicated with *Estr.* for structural parameter or level zero of specification, and Func.x for functional parameter of level x of specification. The functional constants that appear during parameter equation proposal are: $\mathfrak{M}_s$ with $s = A, B; K_{P,c},$ $t_{I,c}$ and $t_{D,c}$ with $c = C, L, T; t_{CA}, k_0, E_a, R, \rho_J, \mu_J, k_m, D_{eq,J},$ and $K_J.$

TABLE 3.3
Constitutive and Assessment Equations for Model Functional Parameter

Parameter	Type	Equation	Instances	Total
$\mathfrak{M}_{mix}$	Estr.	$\mathfrak{M}_{mix} = x_{A,2}\mathfrak{M}_A + x_{B,2}\mathfrak{M}_B + x_{S,2}\mathfrak{M}_S$	...	1
ρ_{mix}	Estr.	$\rho_{mix} = \left(\dfrac{w_{A,mix}}{\rho_A} + \dfrac{w_{B,mix}}{\rho_B} + \dfrac{w_{S,mix}}{\rho_S} \right)^{-1}$	...	1
$w_{j,mix}$	Func.1	$w_{A,mix} = x_{j,mix}\dfrac{\mathfrak{M}_j}{\mathfrak{M}_{mix}}$	$j = A, B, S$	3
$\dot{n}_i$	Estr.	$\dot{n}_i(k) = \dot{n}_i(k-1) + K_{P,i}\left[e_i(k) - e_i(k-1)\right] +$ $\dfrac{K_{P,i}t_{CA,i}}{t_{I,i}}e_i(k) + \dfrac{K_{P,i}t_{D,i}}{t_{CA,i}}[e_i(k) - 2_{ei}(k-1) +$ $e_i(k-2)]$	$i = C, L$	2
$e_i(l)$	Func.1	$e_i(l) = i_{SP}(l) - C(l)$	$i = C, L$ and $l = k,$ $(k-1), (k-2)$	6
i_{SP}	Func.2	$i_{SP} = Datum$	$i = C, L$	1
r	Estr.	$r = k_0 C_{A,2} exp\left(-\dfrac{E_a}{RT}\right)$	...	1
$C_{A,2}$	Func.1	$C_{A,2} = x_{A,2}\dfrac{1}{\rho_{mix}}$	...	1
$\dot{\rho}_{mix}$	Estr.	$\dot{\rho}_{mix} = \dfrac{\dot{m}_1 - \dot{m}_2}{V_I}$	...	1
V_I	Func.1	$V_1 = A_T L$	...	1
N_I	Estr.	$N_I = A_T L \rho_{mix}\dfrac{1}{\mathfrak{M}_{mix}}$	...	1
$x_{A,1}$	Estr.	$x_{A,1} = Datum$	...	1

(Continued)

TABLE 3.3 *(Continued)*

Constitutive and Assessment Equations for Model Functional Parameter

Parameter	Type	Equation	Instances	Total
$\dot{N}$	Estr.	$\dot{N} = \dfrac{A_T}{\mathfrak{M}_{mix}}\left(L\dot{\rho}_{mix} + \rho_{mix}\dot{L}\right)$	...	1
$\dot{L}$	Func.1	$\dot{L} = \dfrac{\mathfrak{M}_{mix}}{\rho_{mix}A_T}\left(\dot{n}_1 - \dot{n}_2 + r\sum_i \sigma_i\right) - \dfrac{L}{\rho_{mix}}\dot{\rho}_{mix}$	...	1
$x_{B,1}$	Estr.	$x_{B,1} = Datum$	...	1
$\bar{C}_{P,mix}$	Estr.	$\bar{C}_{P,mix} = \bar{C}_{P,2}$	...	1
$\bar{C}_{P,i}$	Func.1	$\bar{C}_{P,i} = x_{A,i}\bar{C}_{P,A} + x_{B,i}\bar{C}_{P,B} + x_{S,i}\bar{C}_{P,S}$	$i = 1, 2$	2
$\bar{C}_{P,A}(T_i)$	Func.1	$\bar{C}_{P,A}(T_i) = f_A(T_i)*$	...	1
$\bar{C}_{P,B}(T_i)$	Func.1	$\bar{C}_{P,B}(T_i) = f_B(T_i)*$	...	1
$\bar{C}_{P,S}(T_i)$	Func.1	$\bar{C}_{P,S}(T_i) = f_S(T_i)*$	...	1
$x_{S,i}$	Func.1	$x_{S,i} = 1 - x_{A,2} - x_{B,2}$	$i = 1, 2$	2
$\bar{H}_{mix,i}$	Estr.	$\bar{H}_{mix,i} = \displaystyle\int_{T_{ref}}^{T_i}\bar{C}_{P,i}(T)** \, dT$	$i = 1, 2$	2
$\dot{Q}$	Estr.	$\dot{Q} = U\,A_H\left(T_2 - T_{J,3}\right)$	...	1
U	Func.1	$U = Datum$	...	1
$T_{J,3}$	Func.1	$T_{J,3} = Datum$	...	1
$\dot{m}_3$	Estr.	$\dot{m}_3(k) = \dot{m}_3(k-1) + K_{P,T}\left[e(k) - e_T(k-1)\right]$ $+ \dfrac{K_{P,T}t_{CA,T}}{t_{I,T}}e_T(k) + \dfrac{K_{P,T}t_{D,T}}{t_{CA,T}}[e_T(k) - 2e_T(k-1)$ $+ e_T(k-2)]$	...	1
$e_T(l)$	Func.1	$e_T(l) = T_{SP}(l) - T(l)$	$l = k, (k-1), (k-2)$	3
T_{SP}	Func.2	$T_{SP} = Datum$	...	1
$\hat{C}_{P,J}$	Estr.	$\hat{C}_{P,J} = \dfrac{1}{2}\left(\hat{C}_{P,J,3} + \hat{C}_{P,J,4}\right)$	...	1
$\hat{C}_{P,J,i}$	Func.1	$\hat{C}_{P,J,i} = f_J(T_i)*$	$i = 3, 4$	2
$\hat{H}_{J,i}$	Estr.	$\hat{H}_{J,i} = \displaystyle\int_{T_{ref}}^{T_i}\bar{C}_{P,J,i}(T)** \, dT$	$i = 3, 4$	2
P_3	Estr.	$P_3 = Datum$	...	1
$h_{f-3,4}$	Estr.	$h_{f-3,4} = f\left(K_J, N_{Re,j}\right)$	...	1
$N_{Re,J}$	Func.1	$N_{Re,J} = \dfrac{\rho_J v_J D_{eq,J}}{\mu_J}$	...	1
v_J	Func.2	$v_J = \dfrac{\dot{V}_J}{A_{F,J}}$	...	1
			Total eqs.	**=51**

* Particular polynomial functions of temperature T for each substance.

** $\bar{C}_{P,mix}$ is a polynomial of T obtained from the polynomials of T for the $\bar{C}_P$ of each substance.

Molecular Mass of the Mixture. Using the assumptions of ideal behavior of the mixture and perfect agitation, the constitutive equation for this structural parameter is

$$\mathfrak{M}_{mix} = x_{A,2}\mathfrak{M}_A + x_{B,2}\mathfrak{M}_B + x_{S,2}\mathfrak{M}_S \tag{3.14}$$

from which the molecular mass of *A*, *B*, and *S* are three new functional constants produced by this equation.

Reactor Feed Flow, Reactor Outlet Flow, and Jacket Inlet Flow. This flows are the manipulated variables of the concentration, the level, and the temperature controllers, respectively. As PID controllers are being used for all three control loops, the next generic function in discrete form allows to calculate the value of the flow $\dot{n}_i$ with $i = 1, 2, 3$:

$$\dot{n}_i(k) = \dot{n}_i(k-1) + K_P\left[e(k)-e(k-1)\right] + \frac{K_P t_{CA}}{t_I}e(k) + \frac{K_P t_D}{t_{CA}}\left[e(k) - 2e(k-1) + e(k-2)\right]$$

$$\tag{3.15}$$

from which the controller adjustments K_P, t_I, and t_D are constant values and the errors $e(k)$, $e(k-1)$, and $e(k-2)$ appear as three new functional parameters. These are functional parameters of level 1 of specification defined as

$$e(l) = y_{SP}(l) - y(l) \tag{3.16}$$

with $l = (k)$, $(k-1)$, $(k-2)$, $y(l)$ the value of controlled variable at time l. For the CSTR, y could be concentration C, level L, or temperature T. $y_{SP,v}(l)$ is a new functional parameter called set point or desired value of controlled variable $v = C, L, T$. The value of $y_{SP,v}(l)$, a parameter of level 2 of specification, is given as an operative condition of the process: $y_{SP,v}(k) = Datum$, with values for $v = C, L, T$.

Finally, going back to Eq. 3.15, four new functional constants appear: K_P, t_I, and t_D, the proportional gain, integral time, and derivative time of the PID controller, and t_{CA} the controller action time. These constants are defined as process operative conditions due to their high probability of change during process operation.

Process conditions. These conditions are concentration in the feed stream, the temperature of coolant fluid, and its pressure at jacket input. These parameters are calculated using the next trivial assessment: $x_{A,1} = Datum$, $x_{B,1} = Datum$, $T_{J,3} = Datum$, and $P_3 = Datum$, considering that these data are directly taken from the operating condition of the process.

3.6.8 STEP 8: ACCOUNT OF DEGREE OF FREEDOM

This step prescribes to verify the degrees of freedom (*DF*) of the model structure, i.e., the mathematical systems formed by all model equations (basic structure and extended structure). The *DF* are evaluated as the difference between the number of unknowns and the number of equations. *DF* must be zero for a solvable model. This is a simple count to check if any unknown has an equation to find its value.

For the CSTR using here as case study, the unknowns are the 7 variables, 20 structural parameters, and 31 functional parameters for a total of 58 unknowns. The equations are 7 from model basic structure and 51 for model parameters for a total of 58 equations. Therefore, the degrees of freedom are zero, indicating that current model is solvable. The apparently high number of equations is not a cause for concern because the majority of that equations are very simple assessment equations.

3.6.9 STEP 9: CONSTRUCTION OF A COMPUTATIONAL MODEL

This section discusses ways to build a computational model, i.e., a computer program with the capability of solving the model equations without altering the true mathematical model response. As it was said, the PBSM has basic structure and extended structure. The required computational model translates the total model structure to computational code because this is the solvable one in accordance with previous DF evaluation. Care should be taken not to confuse the PBSM, which is a mathematical model, with the computational model, which is a software procedure. Again, the lumped or distributed parameters characteristic of the model must be considered during computational model construction. Lumped models are commonly used, existing a lot of material about programming procedures. For distributed parameters computational models, a complete methodology could be found in [19].

For the CSTR model treated here, all model equations were programmed and solved using MATLAB®. The obtained code is simple to follow due to lumped parameter characteristic of the developed PBSM.

3.6.10 STEP 10: MODEL VALIDATION

To validate the model response is the last step to finish the modeling procedure. The ideal situation is to contrast model predicted values with real data taken from the process under the operating conditions established in Step 2 for the model objective. If real process data are not available, it is possible to validate the model response against expected behavior of the process, in accordance with available knowledge about the behavior of similar processes. In that sense, the main behaviors of the process must be ever reflected in the model responses.

For the case study, the model validation is inherited from the authors who proposed this CSTR benchmark. By space limitations, parameters values are not reported here, but the reader is invited to review [16]. To show the validity of deduced PBSM, the three PID controllers were tuned by trial and error. Next, step disturbances were applied over inputs at 10 and 180 minutes from simulation beginning. The first step disturbances were all increasing regarding nominal values and the second step disturbances were all decreasing from nominal values: $\pm 0.05 \frac{gmol}{L}$ in input reagent concentration, $\pm 3°C$ in rector feed temperature, and $\pm 2°C$ in coolant inlet temperature. In the following, the Level L response is not presented due to space limitation and its good behavior (only 0.04 m of overshoot and 12 minutes spending to return to set point). In Figure 3.2, the dynamical behavior of reagent concentration is shown at the top and the behavior of concentration control valve at the bottom. In the Figure 3.3, the temperature behavior is shown at the top and the control valve movements at the bottom. This is the kind of model validation possible for known processes without data from a real assembly.

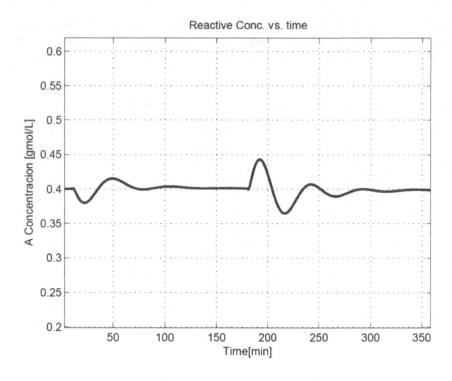

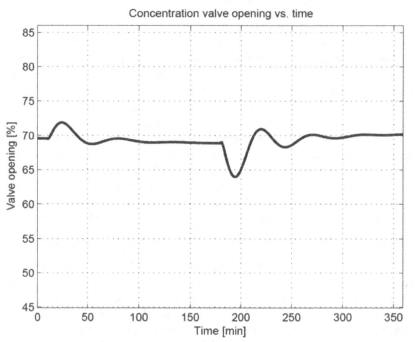

FIGURE 3.2 Concentration dynamic behavior.

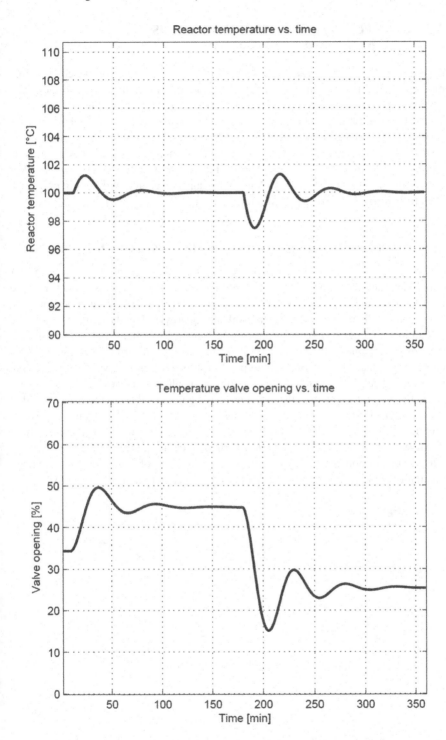

FIGURE 3.3   Temperature dynamic behavior.

3.7 EXAMPLES OF MODELING HYPOTHESIS AND PROCESS PARTITIONS

Considering that Steps 2 and 3 are decisive in the successful of proposed modeling methodology, this section presents three examples of those steps in typical models of process engineering and human physiology. To determine a model hypothesis (Step 2) and to partition the real process in process systems (Step 3), strongly define the final model representation capabilities. To state a suitable modeling hypothesis based on a good process partition, can be done directly from real process behavior or using an analogy. In this way, it is possible to obtaina useful process description. After that, remaining modeling methodology steps can be executed without big problems, even for a novice modeling practitioner.

3.7.1 MODEL OF THE ROLE OF STOMACH IN HUMAN GLUCOSE HOMEOSTASIS

The role of human stomach in glucose homeostasis is represented by this model. A detailed model deduction is presented in [20]. The model hypothesis takes a closed pipe loop as stomach analogy. Through that circuit the food under digestion is in constant movement due to the mechanical energy provided by the stomach wall. The evaluation of required energy to do this work is done through mechanical energy balance over each one of the straight pipe line and fitting conforming the circuit. That energy is produced by glucose consumption of stomach muscle, considering a 50% of efficiency of cell forming that tissue for converting glucose in energy. This glucose is taken from blood stomach irrigation altering the glucose values in the whole blood of the body. In addition, that food movements produce the partial digestion of lipids and proteins into stomach. All these changes are modeled as chemical reactions using Arrhenius law. Due to that digestion, the rheological properties of the food change, altering the mechanical energy exchange. On left side of Figure 3.4, the analogy taken for the model construction is illustrated. The evolution of glucose consumption

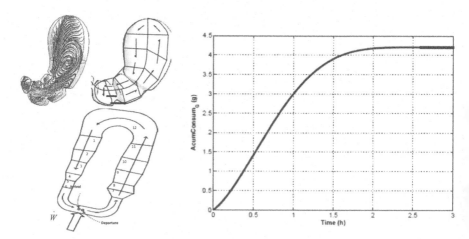

FIGURE 3.4   Analogy and response of the model for stomach role on glucose homeostasis.

due to stomach mechanical work during the stomach digestion process is presented on the right side of Figure 3.4. Other results of the model can be reviewed in [20].

3.7.2 MODEL OF THE ROLE OF LIVER IN HUMAN GLUCOSE HOMEOSTASIS

In the same line that previous example, PBSMs can be developed for human organs involved in glucose homeostasis. The model of the role of liver in that regulation task is discussed in brief in the following, focusing on the modeling hypothesis and partition in process systems. A more wide treatment is presented in [21]. The model is proposed using an analogy of liver functions related to glucose with two agitated tanks, one of them acting as a multiphase reactor and the other as blood container. On the left side of Figure 3.5 that analogy is illustrated. Two process systems are considered: PS_I for the blood contained in the circulatory subsystems irrigating the liver and PS_{II} for the reactive mass contained in the reactor. This reactive mass contains the glucose and glycogen considered immiscible. Glucose and glycogen act as product or reagent. In addition, the reactive mass contains glutamine, glycerol, alanine, and lactate, the non-glucidic glucose precursors. Finally, the insulin and glucagon are considered as part of this reactive mass in PS_{II} due to their activation or inhibition role on liver glucose production or storage. The insulin and glucagon clearance in the liver is modeled as an additional chemical reaction.

The modeling hypothesis used for this model is as follows. The liver takes from its blood irrigation systems the glucose and non-glucidic biochemical precursors to form glucose. A part of the insulin and the glucagon present in the blood are taken too. In accordance with insulin and glucagon concentration in blood, the liver executes two different reactions: *i*) glucose conversion in glycogen, which is a polymer of glucose stored in the liver when glucose level in blood is high; *ii*) glycogen conversion in glucose and conversion of non-glucidic substances in glucose. In that way, the liver tries to maintain the glucose in blood level into the interval $[70,150] \frac{mg}{dL}$, considered as normal for a healthy person. The liver behavior during the sequence fasting, breakfast, and postprandial is illustrated as solid line on the right side of Figure 3.5. The dashed line correspond to a simulated signal of blood glucose

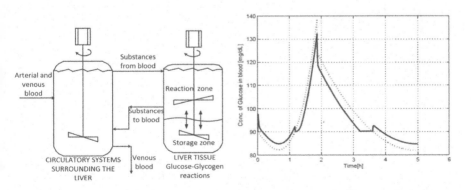

FIGURE 3.5 Analogy and response of the model for liver role on glucose homeostasis.

concentration previous to liver. It should be highlighted that this liver inlet concentration has no real measurements in humans due to technical difficulties and ethical aspects. Expected form of this model input is deduced from measurements in rats and dogs. As the model reproduces the concentration changes of glucose, insulin, glucagon, and non-glucidic precursors, model validation was done using reported measurement in humans for those variables [22, 23]. The maximum error between model and available data for that substance concentrations in blood was less than 10%, a good level for prediction and control purposes. A complete model of the role of liver in human glucose homeostasis can be found in [24].

3.7.3 MODEL OF FRICTION STIR WELDING PROCESS

The process consists of welding two sheets of metal using the technique of stirring by friction, as it is illustrated on the left side of Figure 3.6 and explained in detail in [14]. An analogy was necessary because the real phenomena taken place are currently unknown or are described in specialized literature as very complex.

The used model hypothesis refers to the analogue process illustrated on the right side of Figure 3.6. In brief, the modeling hypothesis is as follows, considering the next four process systems: PS_I the tool as a driving machine, PS_{II} the material between threads and base material wall, PS_{III} a restriction valve, and PS_{IV} the metal into rear cavity formed during the tool advance, which is the desired welding. The hypothesis considers the metal is removed from both sheets and transported into air as pulverized metal. The mechanical energy used to maintain this two-phase flow is provided by a driving machine. That machine is formed in the taken analogy by the tool and the walls of the two sheets. The air-metal mixture passes first through a tube bent as a serpentine, which is the simile of the thread of the welding tool. After that, this fluid crosses a restriction valve before forming the welding at the rear of the tool. The model predicts the level of filling of that consolidated metal. This variable indicates the goodness of the welding. For space limitation, the reader is invited to review the work of [14] for detailed results, indicating that the error of the model predicting the welding goodness is less than 15%.

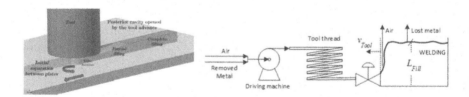

FIGURE 3.6   Friction stir welding process.

REFERENCES

1. R.B. Bird, W.E. Stewart, and E.N. Lightfoot. *Transport Phenomena*. John Wiley & Sons, New York, USA, 1960.
2. R. Aris. *Mathematical Modelling Techniques*. Pitman, London, UK, 1978.

3. L. Ljung. *System Identification: Theory for the User.* Prentice Hall Inc., New Jersey, USA, 1999.
4. P. Humphreys. *Extending Ourselves. Computational Science, Empiricism, and Scientific Method.* 1st ed. Oxford University Press, New York, USA, 2004.
5. S. Filizadeh. *Electric Machines and Drives: Principles, Control, Modeling, and Simulation.* CRC Press, N.J., USA, 2017.
6. F. Bullo, and A.D. Lewis. *Geometric Control of Mechanical Systems: Modeling, Analysis, and Design for Simple Mechanical Control Systems.* Springer, N.Y., USA, 2004.
7. J.L. Diaz, D.A. Mũnoz, and H. Alvarez. Phenomenological based soft sensor for online estimation of slurry rheological properties. *International Journal of Automation and Computing,* 2:1–11, 2018.
8. U. Jordan, and A. Schumpe. The gas density effect on mass transfer in bubble columns with organic liquids. *Chemical Engineering Science,* 56(21):6267–6272, 2001. (Proceedings of the 5th International Conference on Gas-Liquid and Gas-Liquid-Solid Reactor Engineering.)
9. L. Lema-Perez, R. Mũnoz-Tamayo, J.F. Garcia-Tirado, and H. Alvarez. On Parameter interpretability of phenomenological-based semiphysical models. bioRxiv, 2018.
10. D. Basmadjian. *The Art of Modeling in Science and Engineering.* Chapman, New Jersey, USA, 1999.
11. K. Hangos, and I.T. Cameron. *Process Modelling and Model Analysis,* vol. 4. Academic Press, US A, 2001.
12. F.A. Ortega, H. Alvarez, and H. Botero. Enfrentando el modelado de bioprocesos: una revisión de las metodologías de modelado. *Revista Ión. UIS.,* 30(1):73–90, 2017.
13. H. Alvarez, R. Lamanna, P. Vega, and S. Revollar. Metodología para la obtención de modelos semifísicos de base fenomenológica aplicada a una sulfitadora de jugo de caña de azúcar. *Revista Iberoamericana de Automática e Informática Industrial RIAI,* 6(3):10–20, 2009.
14. E. Hoyos, D. López, and H. Alvarez. A phenomenologically based material flow model for friction stir welding. *Materials & Design,* 111:321–330, 2016.
15. C. Zuluaga-Bedoya, M. Ruiz-Botero, M. Ospina-Alarclón, and J. Garcia-Tirado. A dynamical model of an aeration plant for wastewater treatment using a phenomenological based semi-physical modeling methodology. *Computers & Chemical Engineering,* 117:420–432, 2018.
16. S.C. Sherman, A.V. Iretskii, M.G. White, and D.A. Schiraldi. A new stereoselectivity process for the manufacture of dimethylbiphenyl from toluene. *Chemical Innovation,* 30(7):25–30, 2000.
17. P. Atehortúa, H. Alvarez, and S. Orduz. Modeling of growth and sporulation of bacillus thuringiensis in an intermittent fed batch culture with total cell retention. *Bioprocess and Biosystems Engineering,* 30(6):447–456, 2007.
18. S. López, J. García-Tirado, and H. Alvarez. A methodology for identifying phenomenological-based models using a parameter hierarchy. *The Canadian Journal of Chemical Engineering,* 98(1):213–224, 2020.
19. J.F. Wendt. *Computational Fluid Dynamics.* Springer, Berlin, 2009.
20. L. Lema-Perez, J.F. Garcia-Tirado, C. Builes-Montãno, and H. Alvarez. Phenomenological based model of human stomach and its role in glucose metabolism. *Journal of Theoretical Biology,* 460:88–100, 2019.
21. M. Monsalve, and L. Jimenez. Semi-physical model of liver role in the glucose homeostasis. In *2018 AADECA Congress.* AADECA Argentina, 2018 [In Spanish].
22. J.E. Gerich. Physiology of glucose homeostasis. *Diabetes, Obesity and Metabolism,* 2000.

23. M. Konig, S. Bulik, and H. Holzhutter. Quantifying the contribution of the liver to glucose homeostasis. *PLoS Computational Biology*, 8(6), 2012.

24. C.E. Builes-Montaño, L. Lema-Pérez, J. García-Tirado, and H. Alvarez. Main glucose hepatic fluxes in healthy subjects predicted from a phenomenological-based model. *Computers in Biology and Medicine*, 142(1):32–44, 2022.

4 Multiscale Modelling
From Approaches to Membrane Technology Simulations

Francesco Petrosino, Giorgio De Luca,
and Stefano Curcio

4.1 INTRODUCTION

4.1.1 MULTISCALE MODELLING IN CHEMICAL ENGINEERING

Multiscale modelling methods are receiving more interest in several technological and scientific areas. "Multiscale simulation can be defined as the enabling technology of science and engineering that links phenomena, models, and information between various scales of complex systems. The idea of multiscale modeling is straightforward: one computes information at a smaller (finer) scale and passes it to a model at a larger (coarser) scale by leaving out, i.e., coarse-graining, degrees of freedom" [1]. Multiscale modelling is becoming of fundamental importance, for instance, in drug design and is having a very high influence on human health; drug function is developed at the molecular scale but definitely has macroscopic effects, so its behavior has to be analyzed, considering the impact it may have at different time and space scales.

Currently, one of the major chemical engineering challenges is inherent in the description of complex phenomena, which are basically multiscale in nature. Chemical engineering is increasingly trying to describe these complex phenomena by formulating (and solving) multiscale models. Generally speaking, it is possible to distinguish three kinds of multiscale methodologies: *descriptive, correlative,* and *variational.* The *descriptive* approach describes the phenomenological differences of structure at different scales. The *correlative* approach allows explanation of a certain phenomenon at larger time and space scales by analyzing some of the mechanisms characterizing its behavior at smaller scales. The *variational* approach is used to unveil the structure prevalent in mechanisms and the mutual relationships existing among different scales [2].

A possible scheme of these relationships (and the corresponding transitions) that are to be considered to develop a complete description of a chemical engineering phenomenon according to a multiscale approach is shown in Figure 4.1.

A multiscale approach, therefore, requires both a detailed study at each of the scales, which characterizes the system behavior, and the identification of some linking

DOI: 10.1201/9781003435181-4

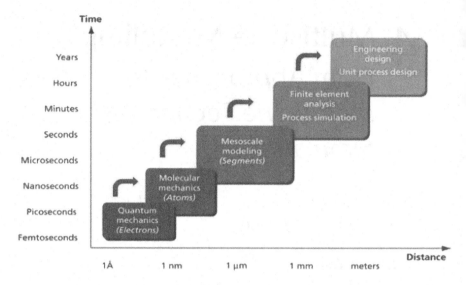

FIGURE 4.1   Characteristics, lengths, and times of multiscale methodologies.

parameters allowing a proper transition (coarse graining) between different scales. No single simulation method exists to attain a detailed insight about real process performance considering their actual multiscale nature; in addition, no general and comprehensive methodology can be defined to suggest how to link and couple the phenomena occurring at different scales. Multiscale modelling, on the other hand, allows properly analyzing and simulating the connections among the multiple scales so to define how a perturbation or even a change at one scale may affect the results at another scale of detail. Nowadays, a challenge is still represented by the proper integration simulations with data across different time and space scales [3].

Seminal contributions to multiscale approach have been made by Karplus, Levitt and Warshel (Nobel Prize in Chemistry, 2013) with the development of multiscale methods for modelling complex biochemical systems [4]. Today, 50 years after their pioneering studies, these methods offer the possibility of analyzing very complex and sophisticated systems. Multiscale modelling provides, for example, different information on how a chemical reaction that takes place in an enzyme active site can influence other proteins and extend from subcellular neighborhoods to cells and tissues through the hierarchy of biological complexity. The advance of multiscale methods consisted of the possibility of creating a fluid knowledge landscape that can encompass different modelling and experimental data on a lot of space and temporal scales. A real outcome of this methodology relies on the proper combination of various approaches, thus connecting the atomic scale of chemistry with the cellular biological function at higher levels to explain the mechanisms of fundamental phenomena. This approach can drive in a circular way to drug design and development [5].

Incorporating the scale-up from one scale to another in a multiscale paradigm is a real challenge concerning same material mechanics appliances. Specified properties at the microscale level have to be integrated directly in the macroscale, avoiding any averaging due to the key role of the model. Different basic approaches have a direct volumetric coupling between the size of the macroscale finite element and the representative volume elements (RVE) at the microscale level have been proposed. In distinction to the previously explained methods (coarse graining or homogenization techniques), this solution, called localization technique, is not truly of the homogenization type. Moreover, it is like a domain decomposition approach in which the lower scale is implanted as a local enhancement.

Nevertheless, multiscale modelling is not constrained to chemical engineering areas: the medical, biological, environmental, and astrophysical fields also use such approaches.

Assuming the value of these modelling techniques and the different studies carried out [6, 7], the present chapter appears to be a fundamental element in the multiscale modelling approaches, which in turn are important in various fieldsranging from bioengineering, up to, process engineering, passing through information and communication technology (ICT), areas of artificial intelligence (AI), and neural networks. Machine learning (ML) and AI have gradually become recognized more as engaging technologies in the biomedical, biological, and behavioral sciences. In the biomedical field, multiscale modelling and ML can reciprocally benefit each other. For example, ML methods have newly infused into composite materials design to enable the homogenization of RVE with neural networks or the solution of high-dimensional partial differential equations using deep learning techniques.

4.1.2 AB INITIO MODELS IN MULTISCALE SIMULATIONS

In the last years, ab initio numerical calculations have become a tool for studying the dynamic, structural, and electronic properties of complex polymers, solids, molecules, and clusters of molecules. Nowadays, different atomic processes in chemical and physical fields may be computer-simulated. In 1998, a noticeable symbol of these developments was seen. The Nobel Prize in Chemistry was awarded to the theoretical physicist W. Kohn, and to the quantum chemist J.A. People [8].

Ab initio simulations have become more important in physics and chemistry during recent years. This was especially possible thanks to the dawn of accessible high-performance computers [9].

A quantum computational approach allows answers to two central questions:

• What is the material electronic structure formed by specified atoms?
• What is its structure at the atomic scale?

The goal of ab initio methods in materials science is to virtually "design" the materials having specified properties by changing their specific electronic structure and atomic bonding. However, to address the questions, a lot of theoretical techniques were developed. To compute structural properties of different materials, the numerical methods can be classified:

1. Empirically (i.e., experimentally determined) parameters-based methods.
2. Empirical quantity prescinding methods.

The first set of methods are frequently called *empirical* or *semi-empirical*, while the latter set are called *ab initio* or *first-principles* methods. Ab initio methods are helpful to calculate the electronic properties of clusters, new materials, or surfaces, and for predicting tendencies on an extensive span of materials [10].

In ab initio quantum mechanics methods, the potential energy between atoms is calculated by resolving the electrons Schrodinger equation of lot of atom cores. Due to the very high degrees of freedom, different theories aimed at obtaining an approximate solution to the Schrodinger equation have been formulated. Some of these theories are the self-consistent field (SCF) theory and the density functional theory (DFT) [10].

By using such a multiscale approach developed on the aforementioned ab initio simulations, all the parameters used at the macroscopic level, such as the mass transport coefficients, can be derived from lower scales regardless of any empirical or adjustable parameters. In the case of molecules, characterized by very complex chemical structures composed of thousands of atoms, the overall behavior can be traced back to the single electrostatic interactions, as calculated at smaller time and space scales.

4.1.3 MONTE CARLO MODELS IN MULTISCALE SIMULATIONS

To make the ab initio knowledge robust, thus allowing the transition to nano- and microscopic scale, it is necessary to exploit complex and well-assessed models, such as the Monte Carlo methods.

The term *Monte Carlo* generally refers to all stochastic methods-based simulations to create new system configurations. In the molecular simulation context, specifically the proteins simulation, Monte Carlo refers to the value of equilibrium configuration sampling. In general, a Monte Carlo simulation keeps as follows: initiating from a starting particles configuration in a system, a Monte Carlo move is done to change the particles configuration and their arrangement. Based on an acceptance criterion, this move is accepted or rejected. The acceptance criterion guarantees that during the simulation, all configurations are sampled from a statistical mechanics ensemble distribution and are sampled by a correct weight. After the rejection or acceptance of the programmed move, a property of interest was calculated, and after many moves, an accurate average of this property is taken. A variety of system equilibrium thermodynamic properties can be calculated by the application of statistical mechanics approaches [11].

The illustrated Monte Carlo simulations don't have the restrictions of solving the Newton's motion equations like the molecular dynamic methods. This independence of the Newton's equations allows for randomness in the suggestion of moves. This generates trial configurations within the chosen statistical mechanics ensemble. Even though these moves might be nontrivial, they can direct to huge speedups of computational times, which in some cases are in the order of magnitude of 10^{10} or more in the equilibrium properties sampling. The modeller usually has great flexibility in the approach to a specific problem because of the combining nature of the Monte Carlo moves. Moreover, Monte Carlo techniques are easily parallelizable and could be used with a large CPU clusters machine.

4.1.4 Multidisciplinary Aspects

Multiscale modelling is characterized by the use of many tools and in-depth knowledge of different complex approaches. For these reasons, it is necessary to underline the multidisciplinary aspect.

While multiscale techniques have long been studied in mathematics, the current efforts are actually dedicated to the use of complex mathematical models in applied sciences and technology, in particular, chemistry, fluid dynamics, material science, engineering, and biology. In these fields, problems are often multiphysical in nature; specifically, the different scale processes are governed by physical laws deriving from different natures: i.e., quantum mechanics at one scale and classical mechanics at another.

Emerging from this deep research, an increasing interest for novel mathematical models and new ways of interacting with mathematics is observed. Research areas belonging to mathematical physics or stochastic methods, which so far have exploited rather "traditional" models and computational techniques, are moving faster and faster towards new scenarios that have been opened, for instance, by multiscale modelling. New questions and challenges are, however, arising; new priorities are to be properly evaluated as a result of the fast evolution in computational techniques.

In the past few years, the life science field has assisted to a rapid change in DNA sequences, genomics, proteomics, gene expression, and metabolomics. The capability of software and computers to manage zettabytes (and beyond) of data and the progression in molecular biology experiments, which origin very high amount of data in the field of protein and metabolism, genome and RNA sequence, protein-protein and protein-DNA interaction, gene expression, 3D structure of protein molecules, and more, represent two main brusque vicissitudes in the analyzed field. Life science and computers have become a balancing bridge to each other. This will origin different science branches like the combination of versatile knowledge that caused the dawn of big data in bioinformatics, computational biology, chemical engineering, and biostatistics [12].

The multiscale approach, aimed at proposing the use of innovative computer techniques to analyze the complex phenomena occurring in chemical engineering problems, is perfectly inserted in the ICT field. The study of the complex interactions among macromolecules relies, for example, on the development of codes based on complex approaches such as Monte Carlo and quantum mechanics techniques. Furthermore, the amount of data processed in the analysis of interactions of cluster molecules reveals the existence of strong analogies with big data structures, typical in computer science.

4.1.5 Ultrafiltration Processes

In the industrial field, membrane applications have been shown to be valuable and economically appealing in a variety of processes, e.g. pharmaceutical, protein recovery, alcohol, and latex recovery [13, 14]. The purification and separation of bioproducts like proteins, protein polysaccharides, hydrolysates, amino acids, and vitamins are very important in the food industry due to the high number of potential applications [15, 16]. In particular, ultrafiltration (UF) is extensively used in the treatment of colloidal dispersions and, due to its respectable performance related to cost and selectivity, has become a standard method in protein recovery [17].

UF has different drawbacks, the main one being the flux decrease in the course of permeation. This problem is due both to concentration polarization (CP), a reversible phenomenon raised to the progress of concentration gradients at the interface membrane/solution, and to fouling of the membrane structure, an irreversible phenomenon, which is due also to deposition of solute molecules on the membrane surface creating a layer of gel or to solute adsorption into the pore structure and on the membrane surface [18].

The interplay among applied transmembrane pressure (TMP) and osmotic pressure generate the CP and the formation of fouling that greatly affect membrane performance. To describe membrane fouling, in literature could be founded four classes of models; each one considers the characters of equally the solute(s) dispersed in the solution to be treated and the membrane. The complete blocking model presumes that the solute(s) particles clog the pore ingresses; the intermediate blocking model is like complete blocking but supposes that some of the particles do close the pores, whereas some others accumulate on the surface. The cake filtration model assumes that particles may collect on the membrane surface, forming a permeable cake of increasing thickness, usually responsible for the permeate flow additional resistance. The standard blocking presupposes that colloids or particles in general might accumulate inside the membrane pore walls, which are usually considered to be straight cylinders [19].

In particular, with reference to a membrane simulation multiscale model, it is necessary to emphasize the fact that the concentrated layer of solute particles developing on the membrane is much less permeable to the solvent (usually water) compared to a solution with bulk concentration. Therefore, to appropriately predict the membrane process behavior, it is necessary to accurately calculate both R_{add}, the permeate flow additional resistance, and the $\Delta\pi$, osmotic pressure due to the solute concentration. The osmotic pressure is calculated at the membrane surface, generating a decrease in the driving pressure equal to $-\Delta\pi$. A multiscale approach could represent the key approach to describe such a complex problem.

4.1.6   A Case Study: Multiscale Modelling to Describe UF Processes

Hereafter, the formulation of a multiscale model to simulate useful macroscopic quantities related to UF processes is explained. This modelling approach provides useful information for different processes and bioprocesses where the intermolecular (noncovalent) interactions drive the entire phenomena. Protein adsorption on surface materials is a principal biological phenomenon in nature. It shows a wide application range in different fields as biosensors, membrane-based processes, biofuel cells, biomaterials, biocatalysis, and protein chromatography [20–22]. Thus, the study of interfacial adsorption behavior of proteins is of excellent practical and theoretical significance. It is worthwhile remarking that ab initio simulation allows the estimation of parameters without the use of any empirical or experimental methodology.

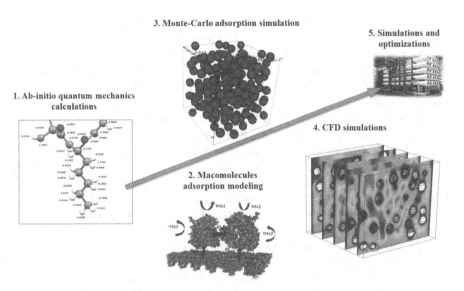

FIGURE 4.2 Multiscale framework of the present work.

Compared to the papers available in the literature [23, 24], the presented multiscale approach contains a major peculiarity. This multiscale analysis was in fact performed on the entire range of possible physical details. It starts from the quantum mechanics scale and reaches the macroscopic scale through a series of coarse graining techniques, allowing the transition between the considered time and space scales (Figure 4.2).

The main objective was to merge and integrate several complex theories in the same multiscale framework and to provide a strong and versatile approach that can be exploited to investigate different areas of interest for science and technology. One of the most relevant challenges is the use of a stochastic approach, which definitely represents a significant breakthrough, since it introduces an improvement in the prediction of the phenomena occurring at nanoscopic and microscopic scale and, therefore, the overall multiscale procedure. The implementation of the Monte Carlo algorithm needs to be performed with the goal of investigating the fouling structure during membrane operation in terms of different micro-equilibrium states. One of the final aims of the work was to achieve the computation of both the osmotic pressure and the diffusion coefficient in the fouling thickness by Monte Carlo simulations. These quantities could be then exploited in macroscopic CFD simulations to calculate the additional resistance to permeation provided by the proteins deposit (cake) accumulated on the surface of the membrane during the UF process.

4.1.6.1 Methodologies

Firstly, based on different works [7, 25–29], an enhanced multiscale modelling intended at describing membrane fouling in the UF process was proposed. A multiscale method to simulate the UF of bovine serum albumin (BSA) aqueous solutions was published beforehand by some of the authors of this chapter [29]. In spite of this, the noncovalent interactions among the membrane surface and proteins were not taken into account in the previous model. Herein, the interactions' protein surfaces were computed accurately by first-principle-based computations taking into account the pH effect. The proteins first layer adsorbed on the membrane and the real surface of polysulfone (PSU) were modeled precisely. The equilibrium distance among adsorbed proteins was computed and fixed as lower-bound to the protein-protein distances in the solid deposit gathered layer on the membrane surface. The calculated BSA surface charges were then used to evaluate the protein potential and the charge density, quantities necessary to define a force balance at a microscopic scale level. The surface potential on the protein was compared to different Z-potential measurements of BSA aqueous solution, and a very good agreement was discovered. Finally, additional resistance of the overall system, due to both the loose and compact layers of the deposit, was described, therefore consenting the final evolution to a macroscopic scale, where a mass transfer unsteady-state model was developed to describe the behavior of a characteristic dead-end UF process. A remarkable agreement between experimental and simulated permeate flux decays was obtained, too.

Secondly, as another important application of the same modelling approach, the adsorption of the phosphotriesterase (PTE) enzyme on the PSU surface membrane was also studied through a double-scale computational method [30]. The surface charges of both the enzyme and the membrane were computed at nanoscopic and sub-nanoscopic scales, whereas adsorption of protein was computed at a bigger scale. The adsorption energies were computed as a function of the surface-enzyme distance and, for every distance, different protein tilting angles were tested to find the highest stable orientation of macromolecules. The modelling approach results were valuable for a detailed insight on the enzyme adhesion and to provide information on the binding site orientations. The computed adsorption energies highly agreed with the data of the literature. In addition, the work showed that the binding site of immobilized PTE was less available with respect to the native (in solution) enzyme due to the polymer surface steric hindrance; hence, an efficiency reduction should be presumed.

Thirdly, a computational study aimed at estimating the osmotic pressure and the diffusion coefficient describing the cake layer accumulated on the membrane surface, i.e. the deposit loose layers, was performed [31]. Using ab initio results obtained at nanoscopic and sub-nanoscopic scales [6], the surface potential of BSA and the minimum equilibrium distance between two proteins were theoretically calculated. A home-made Metropolis Monte Carlo algorithm intended for the simulation of the layer formation during membrane operations was implemented. For the use of the aforementioned Monte Carlo methods, a Yukawa-based system energy computation methodology was also developed [32]. After a complete validation of Monte Carlo methodologies by the hypernetted chain (HNC) theory [33] and through the calculation of radial distribution function (RDF), different Monte Carlo simulations were performed. Finally, in accordance with the approach proposed by Chun and Bowen [34] the diffusion coefficient and the osmotic pressure describing the concentration-polarization profile developing in the membrane filtration process were computed. These results were finally validated both experimentally and by the HNC theory [33] and could be used at macroscopic scale during a UF process simulation.

Finally, different fluid dynamics simulations through Monte Carlo boxes of molecules were carried out. Thanks to the results of the Monte Carlo simulations, various 3D structures were created. These structures represent the deposit layer at different distances from the membrane. With the help of computer-aided tools, these geometries were imported into a computational tool, and the corresponding meshes allowed the performance of micro-fluid dynamics calculations. From these simulations, a series of macroscopic parameters, such as the flow resistance of the deposit layers generally obtained as a result of experiments, were devised [35].

It is very important to emphasize that these works [6, 31, 35–37], although diverse in their subjects, are connected by a fundamental unifying basic strand, which is represented by the multiscale approach for the study of protein-surface and protein-protein interactions, carrying out useful macroscopic quantities. Furthermore, as discussed, this unifying strand has collapsed into a very extensive and promising multidisciplinary field.

4.1.7 CONCLUSION AND FUTURE PERSPECTIVES

A multiscale paradigm reporting an analysis of the interactions among macromolecules as well as between the macromolecules and the surface of a polymeric membrane was illustrated. All the objectives were addressed in multiscale modelling by the complex management of different theories formulated and exploited at several scales of detail. Various fields ranging from bioengineering up to process engineering, passing through ICT, CFD, and biochemistry areas need to be examined. All the related theories, from molecular modelling up to process simulation, need to be suitably mixed and merged to achieve these goals.

Regarding the future perspectives of multiscale simulations in chemical engineering and specially in membrane processes, it is worthwhile to note that an extended ab initio investigation about the effects of different process conditions on the fundamental parameters/inputs controlling the particles interactions will be performed in future research. In particular, some refinements on a quantum mechanics approach should be also considered. At nano- and microscopic scale, a combined Brownian dynamics/Monte Carlo approach represents a very ambitious idea to be implemented to describe more accurately the colloids' aggregation. Moreover, a biased Monte Carlo approach could be considered to improve the sampling performance, thus allowing description of higher macromolecules concentrations. The Monte Carlo/micro-CFD tool could be used to attain a deeper understanding of deposited layers in different membrane process conditions, and the previously proposed Monte Carlo-derived diffusion coefficient could additionally be calculated by micro-CFD scale analysis.

The presented modelling technique points towards a much broader view, providing a stronger and more versatile approach that can be exploited to investigate different fields of interest for science and technology. Moreover, to make this multiscale modelling more widespread in engineering applications, a complete modelling tool could be developed. A user-friendly interface could also make it attractive in industrial process applications. The illustrated approach allowed for a multi-physics description of many scientific and technological areas, such as material science, chemistry, fluid dynamics, biology, and engineering.

REFERENCES

1. M. Fermeglia and S. Pricl, "Multiscale modeling for polymer systems of industrial interest," *Prog Org Coat*, vol. 58, no. 2–3, pp. 187–199, 2007, doi: 10.1016/j.porgcoat.2006.08.028.
2. J. Li and M. Kwauk, "Exploring complex systems in chemical engineering - The multi-scale methodology," *Chem Eng Sci*, vol. 58, no. 3–6, pp. 521–535, 2003, doi: 10.1016/S0009-2509(02)00577-8.
3. R. E. Amaro and A. J. Mulholland, "Biomolecular simulations are essential tools for drug design and development, and for our understanding of the molecular basis of disease", *Nat Rev Chem*, vol. 2, p. 0148, doi: 10.1038/s41570-018-0148.
4. H. Hodak, "The nobel prize in chemistry 2013 for the development of multiscale models of complex chemical systems: A tribute to Martin Karplus, Michael Levitt and Arieh Warshel," *J Mol Biol*, vol. 426, no. 1, pp. 1–3, 2014, doi: 10.1016/j.jmb.2013.10.037.
5. The Royal Swedish Academy of Science, "Development of multiscale models for complex chemical systems," *Scientific Background on the Nobel Prize in Chemistry 2013*, vol. 50005, 2013, https://www.nobelprize.org/uploads/2018/06/advanced-chemistryprize2013.pdf
6. S. Curcio, F. Petrosino, M. Morrone and G. De Luca, "Interactions between proteins and the membrane surface in multiscale modeling of organic fouling," *J Chem Inf Model*, vol. 58, no. 9, pp. 1815–1827, 2018, doi: 10.1021/acs.jcim.8b00298.
7. F. Petrosino, S. Curcio and G. De Luca, "Modellazione multiscala dei processi di separazione di proteine mediante tecnologie a membrana," University of Calabria, 2015.
8. W. Kohn, "Nobel lecture: Electronic structure of matter-wave functions and density functionals," *Rev Modern Phys*, vol. 71, 1253–1266, 1999, http://dx.doi.org/10.1103/RevModPhys.71.1253
9. M. O. Steinhauser, *Computational multiscale modeling of fluids and solids*, vol. 111, no. 479. Springer-Verlag Berlin Heidelberg, 2008. https://doi.org/10.1007/978-3-662-53224-9.
10. M. P. Johansson, V. R. I. Kaila and D. Sundholm, "*Ab initio*, density functional theory, and semi-empirical calculations," *Biomolecular Simulations*, Methods in Molecular Biology, vol. 924, Humana Press, pp. 2–27, 2013, doi: 10.1007/978-1-62703-017-5.
11. A. Kukol, *Molecular modeling of proteins* (Second edition). 2014. doi: 10.1007/978-1-4939-1465-4.
12. R. Martiz, "Application of Big Data Bioinformatics – a Survey," pp. 206–212, 2016.
13. T. W. Cheng, H. M. Yeh and C. T. Gau, "Flux analysis by modified osmotic-pressure model for laminar ultrafiltration of macromolecular solutions," *Sep Purif Technol*, vol. 13, pp. 1–8, 1998, doi: 10.1016/S1383-5866(97)00051-8.
14. S. Chakraborty *et al.*, "Photocatalytic hollow fiber membranes for the degradation of pharmaceutical compounds in wastewater," *J Environ Chem Eng*, vol. 5, pp. 5014–5024, 2017, doi: 10.1016/j.jece.2017.09.038.
15. G. Zin *et al.*, "Fouling control in ultrafiltration of bovine serum albumin and milk by the use of permanent magnetic field," *J Food Eng*, vol. 168, pp. 154–159, 2016, doi: 10.1016/j.jfoodeng.2015.07.033.
16. A. Nath, S. Mondal, S. Chakraborty, C. Bhattacharjee and R. Chowdhury, "Production, purification, characterization, immobilization, and application of β-galactosidase: A review," *Asia-Pac J Chem Eng*, vol. 9, pp. 330–348, 2014, doi: 10.1002/apj.1801.
17. K. Saha, U. Maheswari R, J. Sikder, S. Chakraborty, S. S. da Silva and J. C. dos Santos, "Membranes as a tool to support biorefineries: Applications in enzymatic hydrolysis, fermentation and dehydration for bioethanol production," *Renew Sustain Energy Rev*, vol. 74, pp. 873–890, 2017, doi: 10.1016/j.rser.2017.03.015.

18. A. Suki, A. G. Fane and C. J. D. Fell, "Flux decline in protein ultrafiltration," *J Memb Sci*, vol. 21, pp. 269–283, 1984, doi: 10.1016/S0376-7388(00)80218-5.

19. G. Bolton, D. LaCasse and R. Kuriyel, "Combined models of membrane fouling: Development and application to microfiltration and ultrafiltration of biological fluids," *J Memb Sci*, vol. 277, pp. 75–84, 2006, doi: 10.1016/j.memsci.2004.12.053.

20. X. Quan, J. Liu and J. Zhou, "Multiscale modeling and simulations of protein adsorption: Progresses and perspectives," *Curr Opin Colloid Interface Sci*, vol. 41, pp. 74–85, 2019, doi: 10.1016/j.cocis.2018.12.004.

21. S. A. Bhakta, E. Evans, T. E. Benavidez and C. D. Garcia, "Protein adsorption onto nanomaterials for the development of biosensors and analytical devices: A review," *Anal Chim Acta*, vol. 872, pp. 7–25, 2015, doi: 10.1016/j.aca.2014.10.031.

22. L. Yu, L. Zhang and Y. Sun, "Protein behavior at surfaces: Orientation, conformational transitions and transport," *J Chromatogr A*, vol. 1382, pp. 118–134, 2015, doi: 10.1016/j.chroma.2014.12.087.

23. M. Rabe, D. Verdes and S. Seeger, "Understanding protein adsorption phenomena at solid surfaces," *Adv Colloid Interface Sci*, vol. 162, no. 1–2, pp. 87–106, 2011, doi: 10.1016/j.cis.2010.12.007.

24. L. Zhang and Y. Sun, "Molecular simulation of adsorption and its implications to protein chromatography: A review," *Biochem Eng J*, vol. 48, no. 3, pp. 408–415, 2010, doi: 10.1016/j.bej.2009.12.003.

25. P. Harmant and P. Aimar, "Coagulation of colloids retained by porous wall," *AIChE Journal*, 1996, doi: 10.1002/aic.690421221.

26. M. M. Rohani, A. Mehta and A. L. Zydney, "Development of high performance charged ligands to control protein transport through charge-modified ultrafiltration membranes," *J Memb Sci*, vol. 362, no. 1–2, pp. 434–443, 2010, doi: 10.1016/j.memsci.2010.06.069.

27. Y. S. Polyakov and A. L. Zydney, "Ultrafiltration membrane performance: Effects of pore blockage/constriction," *J Memb Sci*, vol. 434, pp. 106–120, 2013, doi: 10.1016/j.memsci.2013.01.052.

28. S. Curcio, G. De Luca, F. Paone and V. Calabrò, "Modellazione multiscala dei processi di separazione di proteine mediante tecnologie a membrana," 2011. doi: 10.1192/bjp.111.479.1009-a.

29. G. De Luca, F. Bisignano, F. Paone and S. Curcio, "Multi-scale modeling of protein fouling in ultrafiltration process," *J Memb Sci*, vol. 452, pp. 400–414, 2014, doi: 10.1016/j.memsci.2013.09.061.

30. F. Petrosino, S. Curcio, S. Chakraborty and G. de Luca, "Enzyme immobilization on polymer membranes: A quantum and molecular mechanics study," *Computation*, vol. 7, no. 4, p. 56, 2019, doi: 10.3390/computation7040056.

31. F. Petrosino, Y. Hallez, G. de Luca and S. Curcio, "Osmotic pressure and transport coefficient in ultrafiltration: A Monte Carlo study using quantum surface charges," *Chem Eng Sci*, vol. 224, p. 115762, 2020, doi: 10.1016/j.ces.2020.115762.

32. J. P. Edwards, U. Gerber, C. Schubert, M. A. Trejo and A. Weber, "The Yukawa potential: Ground state energy and critical screening," *Prog Theor Exp Phys*, vol. 2017, p. 083A01, 2017, doi: 10.1093/ptep/ptx107.

33. H. L. Lemberg and F. H. Stillinger, "Application of hypernetted-chain integral equations to a central-force model of water," *Mol Phys*, vol. 32, no. 2, pp. 353–362, 1976.

34. M. S. Chun and W. R. Bowen, "Rigorous calculations of linearized Poisson-Boltzmann interaction between dissimilar spherical colloids and osmotic pressure in concentrated dispersions," *J Colloid Interface Sci*, vol. 272, no. 2, pp. 330–339, 2004, https://doi.org/10.1016/j.jcis.2003.12.005.

35. F. Petrosino, G. de Luca, S. Curcio, S. R. Wickramasinghe and S. Chakraborty, "Micro-CFD modelling of ultrafiltration bio-fouling," *Sep Sci Technol*, vol. 58, no. 1, pp. 131–140, 2022, https://doi.org/10.1080/01496395.2022.2075759.

36. F. Petrosino, S. Curcio, S. Chakraborty and G. De Luca, "Enzyme immobilization on polymer membranes: A quantum and molecular mechanics study," *Comput Sci Eng*, vol. 7, no. 4, pp. 1–9, 2019, https://doi.org/10.3390/computation7040056.

37. G. de Luca, F. Petrosino, J. L. di Salvo, S. Chakraborty and S. Curcio, "Advanced descriptors for long-range noncovalent interactions between SARS-CoV-2 spikes and polymer surfaces," *Sep Purif Technol*, vol. 282, 120125, 2021, doi: 10.1016/J. SEPPUR.2021.120125.

5 Modeling Simulation and Chemical Process Optimization

*Radouane El Amri, Reda Elkacmi,
and Otmane Boudouch*

5.1 INTRODUCTION TO MODELING, SIMULATION, AND OPTIMIZATION

Process modeling, simulation, and optimization involve using a model to answer questions about a real-world operation through simulation and optimization. Experiments (simulations) in the virtual world are used instead of costly inquiries in the real world. The model must be predictive in order to serve this objective, i.e., to offer trustworthy answers to problems encountered in the real world. Developing predictive and reliable models is an essential part of the modeling phase (Asprion and Bortz 2018).

Modeling, simulation, and optimization allow chemical process engineers to study the operation of a plant without the need for real equipment, and to avoid the risks of equipment malfunction and environmental impacts encountered in real plants.. It also provides an understanding of the behavior of unit operations and the complex relationships of multicomponent mixtures in reactors and separation units (Ghasem 2018).

The simulation of a chemical process needs the use of a process model, which is a mathematical representation of the process under study. The resulting system of mathematical equations can be either linear or non-linear algebraic equations, or in the form of inequations and partial or ordinary differential equations. These equations are in most cases tricky and complex to solve analytically, but with the development of numerical techniques, modelers have been able to solve the most rigorous problems (Asprion and Bortz 2018).

The mathematical model provides an understanding of the causes and effects of the system being modeled and the complex physical interactions that can occur in the system. Process modeling and simulation are the fastest and most cost-effective techniques for achieving process optimization (Asprion and Bortz 2018).

5.2 CHEMICAL PROCESSES

5.2.1 Definition

Chemical processes are an ordered series of physicochemical transformations that raw materials undergo to produce finished products, marketable by-products, and unwanted waste products.

DOI: 10.1201/9781003435181-5

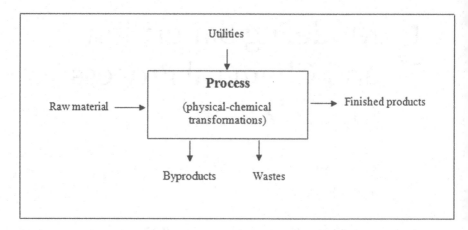

FIGURE 5.1   Typical chemical process scheme.

Chemical processes involve the following phases: storage, raw materials preparation, transformation (reactions), product purification, and packaging. Among these principal phases, there are several different procedures and unitary equipment, such as tanks, agitators, separators, heat exchangers, columns, biological reactors, chemical reactors, pumps, and compressors, among others (Brunet 2014).

Figure 5.1 shows a typical chemical process scheme; the raw materials can be sugar beets, crude oil, coal, air, wastewater, etc. From these raw materials and through several physical and chemical processes, we obtain finished products, valuable by-products, and waste.. These processes require equipment and utilities, which are mainly water, steam, electricity, gas, and refrigerants. For example, from the raw material crude oil, we can obtain naphtha, petrol, diesel fuel, etc. In this refining process, we have equipment such as distillation columns, heat exchangers, ovens, reactors, etc. (Brunet 2014).

5.2.2   CLASSIFICATION OF CHEMICAL PROCESSES

There are many different chemical processes that can be classified according to the compounds to be produced. These processes are mainly found in the oil industry, the pharmaceutical industry, fine chemicals, the plastics industry, the mining industry, the pulp and paper industry, and water treatment (Brunet 2014). Depending on the mode of operation, processes can also be classified as continuous, batch, and semi-continuous processes (Close and Frederick 1995; Felder and Rousseau 2005).

In continuous processes, there are input and output streams. The materials undergoing processing are continuously in movement. Systems that operate continuously work without stopping for real, commercial, and financial purposes. Repeated plant shutdowns affect product quality and can lead to losses or additional production costs. Continuous operation requires process control and automation to control operational parameters such as flows, pressures, temperatures, tank levels, and machine speeds. Examples of continuous processes include wastewater or drinking water treatment, petroleum refining, pulp and paper, natural gas processing, and power generation (Ghasem 2018).

Batch processes are used when small quantities of a chemical are to be produced or only on demand. Pharmaceutical drugs are manufactured by batch processes. A batch process is a system in which the reactor is fed only once with the required amount of material. Once the process is complete, the products are removed from the reactor. The batch process is considered a closed system, as no matter can enter or leave during the operation of the system. Brewing, cooking, and the production of special chemicals are examples of batch processes (Ghasem 2018).

The semi-continuous process is a hybrid of batch and continuous processes. It acts as a continuous process for the incoming streams and as a batch process when the products are withdrawn from the reactor in one go, and vice versa. This process allows the addition of reagents or the elimination of products with time. The flexibility to add reagents all the time with the semi-continuous process represents an advantage compared to the other mode of operation (Ghasem and Henda 2008). It allows for improving the selectivity of the reactions and better controlling the exothermic reactions (Grau et al. 2000).

5.3 MODELING

5.3.1 INTRODUCTION

The model used must be chosen carefully to successfully simulate a chemical process. Modeling, therefore, becomes a necessary precondition for simulation. Developed software and programming techniques have made model construction easy and quick (e.g., object-oriented programming and the graphical user interface). But it still requires a good comprehension of the process, imagination, expertise, and validation of the model.

5.3.2 DEFINITION OF A MODEL

A model is defined as a visual representation of the actual process and can be divided into static and dynamic models. Static models provide a visual image of a system or machine. They are also used to describe and explain the equipment's internal composition. In contrast, dynamic models demonstrate the functioning of a system (Verma 2014).

Process engineers need a quantitative representation of a process. A design that mimics the process is needed, and it is the role of the model to mimic the phenomenon being studied. It can also be mentioned that the procedure of idealizing or approximating physiochemical processes is termed "modelling" (Rice and Do 1995). The model must retain the essential characteristics of the process under study, and the concept must be simple enough to be converted to mathematical equations. If the idealized physical-chemical processes are at the conceptual level, the model can be called a "conceptual model." At this stage, all assumptions and approximations are known. It allows selecting the adequate laws of chemical kinetics, physics, or any other known laws (Verma 2014).

The equations used in the construction of the model are based on the laws of physics. They are used to explain all the interactions and mechanisms involved in a process. For example, the momentum balance, mass balance, and heat balance equations are based on the conservation laws of momentum, mass, and energy, respectively. The set of equations is sometimes called a "mathematical model." The system of equations, if resolved using a numerical technique, can be referred to as a "numerical model." The model equations may also contain some differential equations. In this case, the assumptions are used to define the necessary boundary conditions (Verma 2014).

The model should be suitable for predicting the performance of a process. The different states of a process can be approximated once the model equations are solved. The solution does not necessarily have to predict all aspects of the process. Several models representing only a restricted number of aspects can be put together. Often distinct models are used to describe heat transfer, mass transfer, and reaction kinetics. In addition to transcendental and algebraic equations, the model equations also include differential equations. Therefore, for all these different aspects, the equations of the model are often resolved simultaneously. For example, the model of the evaporation process implies mass and heat transfer equations that are resolved simultaneously (Verma 2014).

Using a model to represent a system, as has been pointed out several times, doesn't reflect all reality. It is limited to some aspects of reality that are considered appropriate to the extent that the model is intended to be applied (Klatt and Marquardt 2009).

5.3.3 Model Limitations

There are numerous variables that do not have a significant influence on the process behavior. Mathematical equations that describe a process with no consideration of these variables can predict the performance of the process adequately. However, variables that can significantly impact the process must still be incorporated into the model. This means that the process can become more responsive to these variables than to others. This information is valuable when developing a control system (Verma 2014).

Although simulation provides a very advanced technique for studying the behavior of all or part of a process, this does not necessarily mean that the resulting information is complete and accurate (Ramirez 1997). The predictions of the studied process are strongly linked to the precision of the applied model. A model is a mathematically idealistic representation of a process and can therefore never give an accurate description of the behavior of the process. Because acceptance of a model requires that it be validated on the basis of experimental data, the model's accuracy also relies on the reliability of the results and the precision of the experimental results. Frequently, when experimental data obtained at the laboratory scale are used for validation of a model constructed for a large-scale system, the results are no longer adequate due to the difference between the two scales. For example, an equation that represents the flow of a fluid in a pipe cannot be extended to represent the flow in a microchannel. The diameter of the microchannel is around a few microns. In the case of a microchannel, the interactions that occur between the molecules of the fluid and the wall of the microchannel are important (Verma 2014).

In the literature, there are several models that give a description of a part of a process. But which of these models can be used, as they are all correct and frequently used by researchers? A model that describes a phenomenon or a process can be based on several theories, so it is necessary to make the right choice among these theories. Consequently, such models are applied in different settings. Careful selection is necessary to provide the most accurate simulated results for the operating range. Because of this, the procedure for developing the simulation is very time-consuming. It includes the time needed to collect data for model validation, to develop the models, and to incorporate smaller models to have a robust model. But once a suitable model is deemed satisfactory for simulating, it may be employed confidently for other comparable applications. This reduces the cost and saves time for the simulation (Verma 2014).

Despite the idealization or simplification of a complicated process through modeling, the model must be capable of capturing the basic characteristics of interest. When these simple models are not able to make feature predictions, a much more advanced model is constructed. An unnecessarily complicated model will only increase the computational effort and will not bring any useful improvement.

5.4 SIMULATION

5.4.1 DEFINITION OF SIMULATION

Simulation is defined as the representation of a physical phenomenon using simple mathematical models to describe its behavior. In other words, simulation allows the various mass, energy, and momentum transfer phenomena that occur in the various unit operations (phase separation, component splitting, compression, expansion, heat exchange, etc.) to be represented by mathematical models. It involves several parameters and the study of their impact on the response of the process. According to Korn (2007), simulation is experimenting with models.

The problems that may be encountered in the simulation of a process are divided into two types. Evaluation problems, in which the system inputs are defined, and the output variables must be found. This type of problem can be solved by directly solving the equations of the model used. For example, in a heat exchanger where a fluid is to be heated, input parameters such as the geometry, flow rate and physical properties of the fluid to be heated, as well as the inlet temperature of the heating liquid, are defined. The model can be used to calculate the flow rate and outlet temperature of the heating fluid.. The second type of constraint is the design problem. In this type, the output parameters are known (desired values), and the input variables that allow having these desired values are to be determined. This form of the problem may be resolved by reversing the model equations (Charpentier 2010). The problem must be treated as an optimization problem. Let's go back to the example of the heat exchanger. If the inlet and outlet temperatures of the process fluid and the flow rate are known, the geometrical parameters such as the number of tubes, the length and the diameter of each tube, the diameter of the shell, the spacing of the baffles, and the flow rate of the heating and cooling fluid must be determined. Several solutions can be obtained. But the solution to be chosen is the most cost-effective one. The model equations are therefore solved based on the optimal parameters (Verma 2014).

5.4.2 Types of Simulation

5.4.2.1 Steady-State Simulation

This type of simulation is applied when the operating parameters of a process do not vary with time (continuous mode). The variation of momentum, mass, and energy with time is zero (zero accumulation). The differential equations thus obtained cannot be derived with respect to time. And consequently, the equations of the model are simple. Nevertheless, this type of simulation cannot predict the behavior in a dynamic process (Verma 2014).

5.4.2.2 Dynamic Simulation

Discontinuous processes are dynamic. The start-up and shut-down of a production or transformation unit that operates in a continuous mode is also a dynamic process. Steady-state simulations cannot study these cases. The dynamic simulation considers time as a parameter of a given process. It allows for determining the optimal operating parameters for a process. For example, the start-up or shut-down time of a process can be optimized. An efficient and economical method of starting and stopping the process may be developed through the use of dynamic simulation (Verma 2014).

Systems based on dynamic simulation can be used for training staff in a production or processing unit. These systems will enable plant operators to master the processes and the different parameters that affect them and to understand the behavior of control systems and unit operations. In addition, this type of training develops the ability of operators to react rapidly when there are changes in system behavior (Verma 2014).

5.4.2.3 Stochastic Simulation

Unlike models that don't take into account the random character of the processes by not implicating random variables (deterministic models), stochastic or Monte Carlo simulations use models that involve random variables of the phenomena, thus allowing the utilization of more fundamental laws of physics (Verma 2014).

5.4.2.4 Discrete-Event Simulation

This type of simulation is used when events occur suddenly at a given time and change the state of the system and when the variables change continuously. For example, conductive heat transfer in a constant temperature roof occurs in a continuous mode. Conduction can be either a steady state or an unsteady state. The fall of particles onto the surface of a fluidized bed is discrete. The addition of reactants to a batch reactor occurs at fixed intervals of time, which is also discrete. The simulation of discrete events is performed with random variables. Likewise, the mathematical description usually implies discrete variables (Verma 2014).

5.4.2.5 Molecular Simulation

Molecular simulation provides an understanding of molecular-scale processes and their relationship to large-scale processes. The models used in this type of simulation consider the molecule's influence on the behavior of a process. Molecular dynamic simulations involving interactions between molecules, movement of molecules, and charges on the walls of the molecule have been performed to predict numerous physical and chemical properties like surface tension, charge of molecules, thermal expansion coefficient, liquid-vapor equilibrium, viscosity, etc. (Gupta and Olson 2003). These simulations allow the study of diffusion phenomena at the nanoscale and the development of pharmaceutical products such as enzymes, biocatalysts, and new drugs. Nevertheless, the simulation of molecular processes has some disadvantages, notably the need for a large amount of computing power and time. Currently, molecular simulations are performed using supercomputers (Verma 2014).

5.4.2.6 Applications of Simulation in Chemical Engineering

Process simulation allows process and chemical engineers to improve an existing process's efficiency, profitability, and design and simulate a new production unit. For example, the simulation may be applied in industrial process design to establish mass and energy balances of the process used and to dimension chemical industry equipment (chemical reactors, heat exchangers, pumps, etc.), or to monitor and control chemical processes and industrial equipment already installed. In the event of a process or equipment modification, the simulation allows to readjust new parameters and operating conditions of the equipment, and to determine the performance of this equipment. Simulations also play a very substantial role in various fields of application, such as process synthesis, equipment design, retrofitting, process control, personnel training, and process safety (Verma 2014).

5.5 OPTIMIZATION

5.5.1 Introduction to Optimization of the Process

Optimization is a technique for determining the most economical and cost-effective solution to a problem or process design. All things in nature, including humans, follow the easy and optimal path to their destination. Humans naturally optimize their efforts in daily life. Heat and water also seek to flow along the path with minimum resistance. Optimization has a long history of use, often based on analytical methods. It has permeated many engineering, science, economics, and business areas. Since the 1960s, several applied mathematical theories for optimization have been elaborated. These theories aim to develop fast and reliable techniques to achieve the optimum of each step of a process. The objective in all areas is to maximize profit and minimize costs. Process engineers are always concerned with managing resources well, maximizing efficiency, and minimizing waste. An optimized process that uses the minimum of inputs (raw materials, energy, personnel, etc.) and provides efficient results in terms of productivity, quality, and environmental friendliness is always beneficial (Edgar et al. 2001).

Chemical industries use many processes and unit operations such as distillation, liquid-liquid extraction, heat transfer, chemical reactors, adsorption, drying, evaporation, etc. Process engineers must ensure that the plant operates under optimal conditions to have maximum profit and minimize losses with minimum environmental impact. Therefore, efficiency is the influencing factor in the optimization of chemical processes (Edgar et al. 2001).

5.5.2 Role of Optimization

Process engineers are always looking for ways to optimize. They work on improving the performance of plants, such as improving the efficiency of equipment and production facilities, achieving more production with lower cost and energy consumption, and obtaining more profit. Optimization also involves the maintenance of equipment and the use of personnel (Edgar et al. 2001).

Modifications made to a production facility or equipment that are intended to be optimized must be carefully considered to see if they will result in benefits. It must be taken into consideration that designs or operating variables are linked in some way. For example, achieving a 5% saving on a $3000 fuel consumption bill in a distillation unit justifies an energy-saving project. However, adding a heat exchanger to a distillation unit to gain sensible heat and reduce the heat required for distillation is incorrect, as reducing the heat load on the boiler can affect the purity of the product, which will result in a variation in profits. Therefore, the indirect effects of process variables on costs should never be overlooked (Edgar et al. 2001).

In some cases, the formal application of optimization may not be justified due to the uncertainty that may exist in the mathematical model that represents the process or uncertainty in the data used. Process engineers must make judgments when applying optimization techniques to problems with significant uncertainties in terms of accuracy or change in plant operating parameters. In some cases, a deterministic optimization analysis can be performed, and then stochastic features are added to the analysis to get quantitative predictions of the level of uncertainty (Edgar et al. 2001).

When the mathematical representation of a process is idealized and the input data and operating parameters are only approximately known, the optimization results must be treated appropriately. They may provide higher limits than expected. Another method to study the influence of uncertain parameters in the optimization of a process is to perform a sensitivity analysis. Some variables in a process have low sensitivity; other parameters do not influence the optimal value of the variable, and therefore, to have the optimum of the process, it is not required to find precise values for these parameters (Edgar et al. 2001).

5.5.3 Classification of Optimization Methods

Optimization techniques can be divided into several categories, depending on the application field, the problem's physical configuration, the design variables' nature, the type of constraints, and the nature of the algorithms. In addition, optimization methods are also classified according to the admissible value of the design variables, the separability of the functions, and the number of objective functions (Dutta 2016).

5.5.3.1 Topological Optimization and Parametric Optimization

Topological optimization is a mathematical method for optimizing the layout of process equipment in a defined design space, for a defined set of loads, constraints, and boundary conditions, with the aim of maximizing the performance of the system. Topological optimization must be considered first, as topological variations often significantly impact the overall performance of the plant. Once the topology of the flow diagram is established, parametric optimization is easy to interpret. Both optimization techniques can be combined and should be used simultaneously (Dutta 2016).

Topological arrangement such as removal of unwanted by-products, equipment rearrangement, alternative reactor configurations, and alternative separation methods can be used to improve heat integration. In contrast, parametric optimization focuses on optimizing operating parameters such as concentration, flow rate, temperature, pressure, etc. (Dutta 2016).

5.5.3.2 Unconstrained and Constrained Optimization

Unconstrained optimization is optimizing an objective function with no additional constraints. Constrained optimization aims to optimize problems where the objective function is performed with a different correlation (constrained function) (Dutta 2016).

5.5.3.3 Linear and Nonlinear Programming

Depending on the structure of the equations used in an optimization problem, it is possible to differentiate between linear and non-linear programming. When all the equations involved in a problem of optimization are linear (objective function and constraint functions), it is referred to as linear programming. On the other hand, when one of these functions is non-linear, this optimization classification is called non-linear programming (Dutta 2016).

5.5.3.4 Convex Function and Concave Function

A function is convex on an interval if its graphical representation on this interval is entirely above each of its tangents. It is demonstrated that a function is convex on an interval if and only if its derivative is increasing on this interval – in other words, if its second derivative is positive on this interval. A function is concave on an interval if its graphical representation on this interval is entirely below each of its tangents. It is demonstrated that a function is concave on an interval if and only if its derivative decreases on this interval – in other words, if its second derivative is negative on this interval (Dutta 2016).

5.5.3.5 Continuous and Discrete Optimization

Continuous optimization is when the objective and constraint functions are continuous, and the values of the decision variables that appear in a problem of optimization are all real numbers. However, when the decision variables are discrete numbers, the optimization is called discrete optimization. A continuous variable is a variable that can take all possible values of an interval of real numbers (time to complete a task, concentration, temperature, pressure, etc.). On the other hand, a discrete variable is a variable that can take only certain values of an interval of real numbers. Generally, the admissible values are only integers (number of workers in a plant, number of plates for the distillation column, number of lots per day, etc.) (Dutta 2016).

In some optimization problems, the objective function contains both integer and real variables; in this case, it is referred to as mixed-integer optimization, another category of the optimization method. For example, the optimization of the operation of a distillation column involves mixed variables such as the temperature and the number of trays (Dutta 2016).

5.5.3.6 Single-Objective and Multi-Objective Optimization

Single-objective optimization is when the problem to be optimized contains only one objective function. This type of problem is easily resolved. Nevertheless, problems that exist in chemical engineering contain more than one objective function (e.g., optimization of the yield and selectivity of a chemical reactor). These optimization problems must be treated carefully, because the objectives are contradictory. Multi-objective optimization is when the optimization problem includes more than one objective function (Dutta 2016).

5.5.3.7 Local and Global Search Method

The local search method consists of finding the optimal solution for a specific region of the search space or the global optimum for problems without local optima. On the other hand, the global search method consists of finding the optimal solution for problems containing local optima. Global optimization methods are generally divided into deterministic and stochastic methods (Dutta 2016).

5.5.3.8 Deterministic, Stochastic, and Combinatorial Optimization

Optimization methods can be classified into two methods, deterministic methods and stochastic methods. Deterministic techniques involve a simple algorithm, which starts from an initial estimate and iteratively updates until a solution is reached. Such optimization techniques generally need knowledge of the form of the objective function. The steepest descent method and Powell's method (conjugate gradient method) are examples of deterministic methods. Deterministic optimization methods have an advantage and a disadvantage. The advantage is that they converge very quickly and with high accuracy by ensuring a reasonable initial estimate of an optimal solution. Their disadvantage is that they tend to lock in local minima when the initial estimate is distant from the optimal solution. Stochastic optimization techniques are based on randomness and repeated trials to obtain a representative sample of the parameter space in the search for an optimal solution. The search point for such algorithms moves randomly within the search space. The frequently used stochastic methods are Monte Carlo, genetic algorithm, and simulated annealing. Stochastic optimization problems are also employed for optimization that seeks to model a process containing data with uncertainties, considering that the input variables are defined according to the probability distribution (Dutta 2016).

Combinatorial optimization is a branch of optimization in the fields of theoretical computer science and applied mathematics. This method is concerned with finding the optimal object among a finite (but often very large) number of objects. A complete search is not possible for this kind of problem. Combinatorial optimization is applied in these problem areas, where all possible solutions are either discrete or convertible to discrete and where the objective is to find the better solution. The most commonly used example of combinatorial optimization is the traveling salesman problem (Dutta 2016).

5.5.4 Applications of Optimization in Chemical Processes

In chemical engineering, optimization can be used in many ways.

Examples of typical optimization projects in chemical engineering are (Edgar et al. 2001):

1. Determination of sites for the establishment of a plant.
2. Planning of distribution truck routes for raw and finished products.
3. Pipe sizing and installation.
4. Plant and equipment design.
5. Planning of maintenance operations and equipment replacement.
6. Operation equipment such as chemical reactors, heat exchangers, and distillation columns.
7. Evaluation of plant data for process modeling.
8. Distribution of resources and services between different processes.

Examples of applications in which optimization is used are as follows (Dutta 2016):

1. Optimization of treated water storage tanks.
2. Optimization of water pumping networks.
3. Optimization of chemical reactors
4. Optimization of distillation columns.
5. Optimization of heat exchangers.
6. Application in thermodynamics and determination of chemical equilibrium.
7. Optimizing a biological wastewater treatment plant.
8. Optimization of drying processes.
9. Inventory management.
10. Design of a real-time controller.
11. PID controller adjustment.

5.6 SIMULATION AND MODELING TOOLS AND SOFTWARE

There are several chemical process simulation and optimization software packages on the market. The following is a non-exhaustive list of the most commonly used software packages worldwide (Brunet 2014):

- Aspen Plus (Aspen Technologies, USA).
- HYSYS (Aspen Technologies, USA).
- SuperPro Desginer (Intelligen, USA).
- Chemcad (Chemistations, USA).
- Pro II (Simulation Sciences, USA).
- Prosim (ProSim, France).
- Ideas (Andritz, Austria).
- SIM42 (France).
- Indiss Plus (Corys, France).
- GAMS (GAMS, USA).

There are several criteria for choosing which simulation software to use, including cost, training of the personnel who will use the software, and adaptability; the software must be well-adapted to the objectives sought.

REFERENCES

Asprion, N., and Bortz, M. 2018. Process modeling, simulation and optimization: From single solutions to a multitude of solutions to support decision making. *Chem. Ingen. Tech.*, 90(11): 1727–1738.

Brunet, R. 2014. *Optimal Design of Sustainable Chemical Processes*. Saarbrücken, Germany: Lap Lambert Academic Publishing GmbH.

Charpentier, J.C. 2010. Among the trends for a modern chemical engineering, the third paradigm: The time and length multiscale approach as an efficient tool for process intensification and product design and engineering. *Chem. Eng. Res. Design.* 88: 248–254.

Close, C.M., and Frederick, D.K. 1995. *Modeling and Analysis of Dynamic Systems*, 2nd ed. New York: John Wiley.

Dutta, S. 2016. *Optimization in Chemical Engineering*. England: Cambridge University Press.

Edgar, T.F., Himmelblau, D.M., and Lasdon, L.S. 2001. *Optimization of Chemical Processes*, 2nd ed. New York City: McGraw-Hill.

Felder, R.M., and Rousseau, R.W. 2005. *Elementary Principles of Chemical Processes*, 3rd ed. Hoboken, NJ: John Wiley.

Ghasem, N. 2018. *Modeling and Simulation of Chemical Process Systems*, 1st ed. Florida: CRC Press.

Ghasem, N., and Henda, R. 2008. *Principles of Chemical Engineering Processes*, 2nd ed. New York: CRC Press.

Grau, M., Nougues, J., and Puigjaner, L. 2000. Batch and semibatch reactor performance for an exothermic reaction. *Chem. Eng. Proc.*, 39(2): 141–148.

Gupta, S., and Olson, J.D. 2003. Industrial needs in physical properties. *Ind. Eng. Chem. Res.* 42: 6359–6374.

Klatt, K.U., and Marquardt, W. 2009. Perspectives for process systems engineering. Personal views from academia and industry. *Comput. Chem. Eng.* 33: 536–550.

Korn, G.A. 2007. *Advanced Dynamic-State Simulation*. Hoboken, NJ: John Wiley.

Ramirez, W.F. 1997. *Computational Methods for Process Simulation*, 2nd ed. Oxford: Butterworth-Heinemann.

Rice, R.G., and Do, D.D. 1995. *Applied Mathematics and Modeling for Chemical Engineers*. New York: John Wiley.

Verma, A.K. (2014). *Process Modelling and Simulation in Chemical, Biochemical and Environmental Engineering*. Florida: CRC Press.

6 An Enhanced Agent-Based Model of the Photocatalytic Hydrogen Production Reaction

Roy Vincent L. Canseco, Rizalinda L. de Leon, Vena Pearl Bongolan, and Joseph Yap IV

6.1 INTRODUCTION

Sustainable and clean energy research is relatively new but highly important. One of the cleanest, most promising and underdeveloped of these technologies is the utilization of hydrogen energy. Both ultraviolet light and visible light have been shown to be able to split water to produce hydrogen in the presence of certain transition metal catalysts such as titania. Visible light accounts for around 50% of solar radiation energy compared to the 3–4% energy in ultraviolet light [1]. Visible light-driven hydrogen production is receiving a lot of attention, with recent research trying to lower costs and improve efficiencies [2, 3]. Photocatalytic experimentation in this field necessitates building reactors and using costly measuring equipment. Each experimental run may also take days to complete. Building a model that readily analyzes current available data will allow for simulations that would suggest the best scenarios to cover in succeeding experiments. A computational model can make the most of the results and new knowledge from each experiment by programmatically integrating it with the existing body of knowledge in the model.

We can identify four steps in the development of new photocatalytic hydrogen reaction processes. The sequence starts with the catalyst identification that involves choosing the appropriate catalyst, with or without dopants, in order to achieve a target energy band gap for the dissociation of water into hydrogen and oxygen. The next step is reagent identification. In this phase, the researchers look at the known mechanisms and see what additional reagents will be needed for the reactions. The next step is concentration optimization. Here experiments are done across different values of reagents in order to determine the best combination of concentrations to use. Finally, testing and validation is done through running experiments using the optimized reagent concentrations.

Each powdered catalyst particle is comprised of free active sites on the surface that bind with reactants and mediate reactions crucial to hydrogen production. It will also bind with reagents, even when such a reaction competes with processes for hydrogen production. Such unwanted filling of catalyst active sites causes catalyst poisoning, which causes the rate of hydrolysis to slow or stop altogether.

DOI: 10.1201/9781003435181-6

As powder in a continuously stirred tank reactor, the cadmium zinc sulfide catalyst are dispersed opaque solids in the solution. Increasing the catalyst, in theory, increases the rate of hydrolysis by providing more free catalyst active sites for the reactions. However, should the amount of catalysts be so dense that the particles near the light source effectively block light from reaching the catalyst particles located deeper in the reactor, the hydrogen production rate plateaus. Any further increase in catalysts causes declines in the production rate [4, 5].

Any reaction product or intermediate that is opaque or translucent also has effects in scattering and blocking light. They compete with the catalyst in the necessary step of photon absorption.

The effects of dense catalyst shadowing, catalyst poisoning, and photon absorption of colored products and intermediates add to the difficulty of assessing heterogeneous reaction rate constants experimentally, making photocatalytic hydrogen reactions unfit for purely mathematical modeling. In light of this, in silico models can help narrow the parameter space. Computer simulations can also explain certain phenomena at the systems level and simulate, to a degree, laboratory experiments. In silico models can be categorized into ab initio (theoretical) and statistical models. Theoretical models are based on established laws and domain knowledge known as first principles. Statistical models are instead derived from observations and data points without a detailed description of the system [6]. Such computational models capture emergent phenomena and provide the flexibility of easily adding new knowledge into the model.

Rule-based modeling is an approach that has successfully been used to address the combinatorial complexity in particular chemical systems [7, 8]. Instead of enumerating all possible reactions that can occur for every spatiotemporal state that can exist in the reactor, a virtual reactor is programmed with rules defining the actions available to a chemical species and the interactions it can have with other species and with the reactor environment. One possible use of the virtual reactor is to find an optimal concentration of reagents which can constitute an experiment design ready for validation.

We developed a simulator named PHyRe v1.3 based on the agent-based model (ABM) paradigm. ABM belongs to a class of discrete models based on stochastic cellular automata programmed on a two-dimensional grid. The entities we represent are representative amounts of chemical species as well as electrons, holes, photons and catalyst active sites. Entities interact when they are on the same patch in the grid (i.e., within the same unit volume) according to reaction procedures, depending on what type of entities are colocated and the type of reaction listed in the mechanism in Section 6.2.1. Entities bind, dissociate, get consumed or get produced during the interaction with other entities. The randomness in the initial placement of the entities on the grid, the programming of a random walk in the movement of the entities and the probability distribution used in determining if a reaction proceeds given all necessary entities share the same coordinates make the model stochastic. Each timestep corresponds to several seconds in real life and with the use of modern quad-core computers, we can simulate in one hour an experiment design that will require a day and for a small fraction of the cost of an actual laboratory experiment. The agent-based platform used is NetLogo [9]. The code has been used for modeling photocatalystic hydrogen production reactions using titanate, cadmium sulfide and cadmium zinc sulfide catalysts [10, 11].

The software allows efficient qualitative parameter sensitivity analysis through reaction rate sliders that can be initialized before the simulation or be set during run-time. PHyRe keeps rules for each reaction in a separate procedure in a way that additional reactions may efficiently be added to the model as new information is discovered. Simulation data can be exported to CSV format and processed by included MATLAB/Octave scripts. Furthermore, MATLAB/Octave functions and scripts were created to take the average between the stochastic batch runs, consolidate simulation results and provide visualization. The scripts are used in qualitative and quantitative assessment of agreement between experimental and simulation results.

The rest of the paper is organized as follows: Section 6.2 describes the chemical basis for conceptual model. Section 6.3 describes the implementation, calibration and compensation of the computational model. Section 6.4 discusses the results of validating, experimentally, what the simulation predicts to be an optimized version of a previously published experiment [4].

6.2 THE CHEMICAL SCENARIO

The photocatalytic hydrogen production on semiconductor particles starts with the absorption of photons with energy higher than the catalyst band gap, thereby generating electron-hole pairs. It mainly involves the reduction of water into hydrogen and the oxidation of sacrificial species on the surface of the catalyst. The recombination of electrons and holes is a competing step. The catalyst, at high concentrations, blocks light and lowers its effectiveness. Desorbed opaque reaction products scatter light and compete with the photon absorption of the catalyst. Additional reactions are engineered to turn such opaque species colorless. Reagents also bind with the catalyst and lower the available active sites, resulting in catalyst poisoning. Section 6.2.1 shows the reaction mechanism for hydrogen production on a cadmium zinc sulfide catalyst [4, 5]. For brevity, the reactions are assumed to take place among adsorbed species on the catalyst. The $Cd_xZn_{1-x}S$ photocatalyst site is represented by Cat.

6.2.1 REACTION MECHANISM FOR HYDROGEN PRODUCTION FROM WATER WITH SULFIDE AND SULFITE IONS

Absorption of photon and generation of electron-hole pair

$$Cat + hv \rightarrow Cat + e^- + h^+ \tag{6.1}$$

Recombination of electron and hole

$$e^- + h^+ \rightarrow \varphi \tag{6.2}$$

Adsorption of reactants

$$Cat + H_2O \rightarrow Cat \cdots H_2O \tag{6.3}$$

$$S^{2-} + H_2O \rightarrow HS^- + OH^- \tag{6.4}$$

$$Cat + HS^- \rightarrow Cat \cdots HS^- \qquad (6.5)$$

$$Cat + SO_3^{2-} \rightarrow Cat \cdots SO_3^{2-} \qquad (6.6)$$

Reactions on the catalyst surface
Reduction of water to hydrogen

$$Cat \cdots H_2O + Cat + e^- \rightarrow Cat \cdots H^{\bullet} + Cat \cdots OH^- \qquad (6.7)$$

$$2Cat \cdots H^{\bullet} \rightarrow 2Cat + H_2 \qquad (6.8)$$

Oxidation of sulfide

$$Cat \cdots HS^- + h^+ \rightarrow Cat \cdots HS^{\bullet} \qquad (6.9)$$

$$Cat \cdots HS^{\bullet} + Cat \cdots OH^- \rightarrow Cat + Cat \cdots S^{\bullet-} + H_2O \qquad (6.10)$$

$$Cat \cdots HS^{\bullet} + Cat \cdots S^{\bullet-} \rightarrow Cat \cdots HS_2^- + Cat \qquad (6.11)$$

Desorption of products

$$Cat \cdots HS_2^- \rightarrow Cat + HS_2^- \qquad (6.12)$$

Reaction of disulfanide with sulfite in the liquid phase

$$HS_2^- + SO_3^{2-} \rightarrow S_2O_3^{2-} + HS^- \qquad (6.13)$$

Adsorption of other species in the system

$$Cat + OH^- \rightarrow Cat \cdots OH^- \qquad (6.14)$$

$$Cat + S_2O_3^{2-} \rightarrow Cat \cdots S_2O_3^{2-} \qquad (6.15)$$

6.2.2 THE CONCEPTUAL MODEL

To describe the previous scenario, one needs to include all the relevant entities that chemical engineers and chemists recognize as relevant in the production process. These entities are the molecules, photons and catalyst active sites. The model is conceptually shown as top-down reaction latter diagrams.

Top-down reaction ladder diagrams show how a chemical species of interest, such as water, becomes hydrogen through a step-by-step series of transformations where the reactants on a higher ladder level transform into products in the level right below it. It tracks a particular chemical species, an entity we shall call 1st reactant, at every rung of the ladder and shows what other entities combine with it in order to move the series of reactions further down the ladder rungs. Having a single chemical entity to track for every reaction/ladder rung is important in agent-based modeling, as interactions always happen through the perspective of an agent or entity.

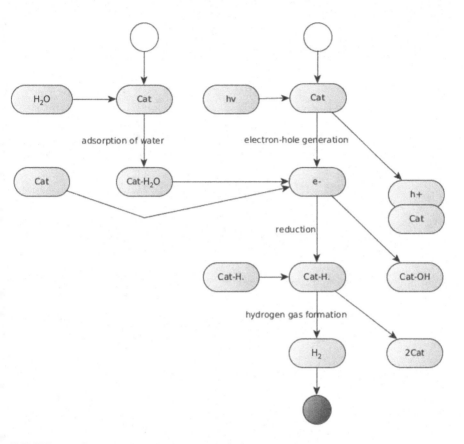

FIGURE 6.1 Conceptual model for water to hydrogen conversion.

We consider top-down reaction ladder diagrams for water to hydrogen conversion in Figure 6.1, for electron-hole recombination in Figure 6.2, for holes scavenging leading to polysulfide formation in Figure 6.3, for the elimination of colored polysulfide products in Figure 6.4 and for the adsorption of various reactants, which compete with hydrogen production in the use of catalyst active sites, in Figure 6.5.

The conceptual model for converting water into hydrogen is shown in Figure 6.1. It contains mechanism reactions 1, 3, 7, 8. The entities involved are water, catalyst

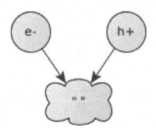

FIGURE 6.2 Conceptual model for the electron-hole recombination.

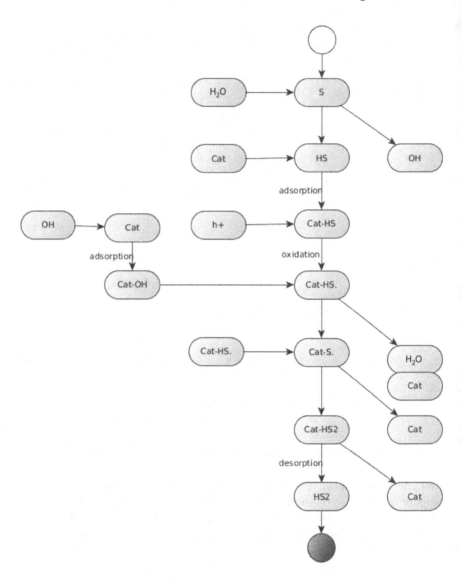

FIGURE 6.3   Conceptual model for the hole scavenging that leads to polydisulfide formation.

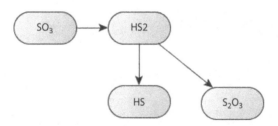

FIGURE 6.4   Conceptual model for the elimination of colored polysulfide products.

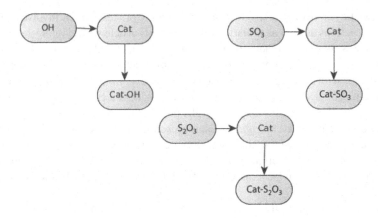

FIGURE 6.5 Conceptual model for the adsorption of reactants contributory to catalyst poisoning.

active sites (Cat), photons (hv), electrons (e–), holes (h+), catalyst bound water molecules (Cat-H$_2$O), catalyst-bound hydrogen radicals (Cat-H.), bound hydroxide ions (Cat-OH) and hydrogen gas.

The conceptual model for the recombination of electron and hole, where both entities cancel each other and disappear, is shown in Figure 6.2. It contains reaction 2 from the mechanism. The entities involved are electrons and holes.

The conceptual model for hole scavenging is shown in Figure 6.3. The holes are removed to reduce and prevent the recombination with electrons, thereby freeing the electrons to further hydrogen production by reacting with bound water molecules. Hole removal is via reactions with catalyst-bound hydrosulfide ions that eventually lead to the production of polysulfides. Polysulfides are yellow-colored products which compete with the light absorption of the catalyst, indirectly lowering electron-hole generation. The mechanism reactions contained in the conceptual model are reactions 4, 5, 9, 10, 11, 12, 14. The entities involved are water, sulfides, catalyst active sites, hydrosulfide ions, hydroxide ions, bound hydrosulfide ions, holes, bound hydroxide ions, bound hydrosulfide ions, bound sulfanyl radicals, bound sulfide radicals, water, catalyst active sites and polysulfides.

The conceptual model for the elimination of colored polysulfide products is shown in Figure 6.4. Polysulfides are colored and would absorb and scatter photons, thereby reducing the available photons for use in electron-hole generation. A reaction with sulfite produces hydrosulfide and thiosulfate, which are both colorless. Reaction 13 is contained in this model. The entities involved are sulfites, disulfanides, hydrosulfides and thiosulfates.

The conceptual model for adsorption reactions contributing to catalyst poisoning are shown in Figure 6.5. For the most part, such reactions lessen the available catalyst active sites that may be used for either hydrogen formation or hole removal. As such, their concentrations should be controlled if possible. Reactions 6, 14 and 15 are contained in this model. The entities involved are catalyst active sites as well as bound and free versions of hydroxide ions, sulfite ions and thiosulfate ions.

Photocatalytic hydrogen production is a complex process which involves hetero-geneous reactions as well as radiation effects. In this article we are interested only in the processes that allow reactant concentration to affect and control hydrogen production in an isothermic continuously stirred tank reactor. The processes happen in a batch reactor with evolved hydrogen gas measured every hour for the duration of the experiment. The actual model, however, only considers a cross section in a virtual region where all interactions take place. The physical reactor is therefore represented by a 2D domain bounded by rigid walls, representing the glass sides of the reactor.

We used an agent-based modeling technique which allows description of the system entities and the interactions between them in a defined space. The system evolution in space and time is generated from the interactions and stochastic behavior of the different entities.

A major advantage of this technique is that the entities and relationships can be described in terms of simple micro-mechanistic behavior, with the bulk mechanics naturally showing in the outcome. The intrinsic non-linearity of the system is treated with no additional effort. The approach is thus more easily understandable and communicable. Another major advantage of this technique is that it is flexible and extensible. The behavior of the entities is modeled based on up-to-date chemical knowledge and may be easily modified to reflect observations from laboratory experiments. Compared to the complexity of real photocatalytic reactions, our model may still be very naïve and can be improved in many aspects. However, the model is sufficiently complete to describe and simulate major aspects of hydrogen production using $Cd_{0.4}Zn_{0.6}S$ catalyst in the presence of sulfide and sulfite ions as sacrificial reagents, which are reagents not directly involved in creating the desired product, but are necessary to prevent other reactions that significantly impede the creation of the product of interest.

6.3   PHyRe 1.3 IMPLEMENTATION

Rule-based modeling is an approach that has successfully been used to address the combinatorial complexity in particular chemical systems [7, 8]. Instead of enumerating all possible reactions that can occur for every spatio-temporal state that can exist in the reactor, a rule-based model defines the actions available to a chemical species and the interactions it can have with other species and with the reactor environment. PHyRe is a NetLogo software model for simulating photocatalytic hydrogen production reactions. The base code has been used for simulating reactions based on titanate, cadmium sulfide and cadmium zinc sulfide catalysts [10, 11].

The simulator takes care of the entity-to-entity interactions that happen during each chemical reaction. The algorithm for an interaction is given by Listing 1. Step 1 makes the computer go through each of the programmed reactions. For each reaction, it picks out the first reactant in the reaction and searches all the agents with same breed (species) as that first reactant. This first reactant is arbitrary and is set to be the first reactant listed in the reactions in mechanism found in Section 6.2.1. The program then asks all the first reactant agents to check their immediate vicinity (i.e. same patch coordinates) for other reactants required in the particular reaction.

In step 2, should agents in step 1 find a complete set of required reactants present in its location, it calls a binary-valued probability distribution function that determines if the reaction will occur during that timestep or if the different agents will simply continue their motion in the reactor environment. The probability distribution uses a parameter equal to the assumed rate constant of the particular reaction. Step 3 assumes the reaction will occur and transforms the reactants into products by initializing new product agents and deleting the reactant agents in the computer simulation. Listing 1 provides a summary of the steps mentioned.

Listing 1: Algorithm for the reactions

1. For each agent with breed = 1st reactant, check its patch for the presence of all other reactants.
2. If required reactants are all present, check the random variable if it is greater than the threshold value (i.e., if the reaction proceeds).
3. If the random variable is greater than the threshold value, hatch the products and kill the reactants.

A feature of these photocatalytic hydrogen production reactions that makes it difficult to model the system as differential equations is the lack of data on the rates of reactions. The rate limiting step is commonly estimated based on the bulk rate diffusion to the surface of the solid catalyst. However, in this case, the catalyst is also lowering the activation energy of the reaction. Since the photocatalytic production of hydrogen in activation is controlled, the problem of determining rates is more complex. Another source of variability is the polydispersed nature of the system. There is a range of catalyst surface reaction rates dependent on the distribution of particle sizes. To allow efficient control over reaction rate constant parameters in the software, graphical sliders corresponding to constant parameters were created that can be initialized before the simulation or be set during run-time.

The list of reactions in the proposed mechanism does not include information on almost all backward reactions, information which can be used to establish expectations on the equilibria [4, 5, 12]. PHyRe keeps rules for each reaction in a separate procedure in a way that additional reactions may efficiently be added to the model as new information is discovered. Simulation data that can be exported to CSV format and processed by included MATLAB/Octave scripts. Furthermore, MATLAB/Octave functions and scripts were created to take the average between the stochastic batch runs, consolidate simulation results, and provide plotting/visualization support. The scripts are used in qualitative and quantitative assessment of agreement between experimental and simulation results.

Taken together, the PHyRe software model and the accompanying MATLAB/Octave scripts provide enhanced capabilities for the photocatalytic hydrogen production simulation as shown in Figure 6.6. Until recently, however, there were several significant shortcomings revealed in newer experiments: (1) large increases the amount of catalyst do not inhibit hydrogen production in simulation as they do in the

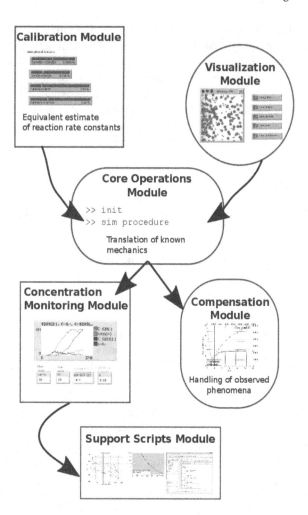

FIGURE 6.6   Basic overview of the PHyRe software.

experiment; (2) the relative speed of electron-hole recombination reactions in simulations was put at maximum rate, but experiments suggest it was still too slow; (4) the equilibria for sulfite is not being reached; and (4) the catalyst active sites are almost never freed once they are bound to any chemical species. Each of these shortcomings has been addressed in the recent PHyRe 1.3 version.

6.3.1 The Catalyst Shadowing Effect

As the amount of catalyst inside the reactor is increased, the light attenuation towards the end of the reactor opposite the lamp also increases. The turbidity of the suspension keeps an amount of light energy from penetrating farther into the reactor. This is negligible for low levels of catalyst and very significant after a threshold is reached [5]. However, the specific manner of how this occurs is not yet clear and is also a matter for research [4]. The uniform stirring action of the magnetic stirrer at 360 rotations per minute prevents the catalyst powder from precipitating. Any variation in catalyst powder sizes is a limitation of the catalyst powder preparation. A regression equation from the available data is programmed into the model to reduce the amount of light.

Agent-based reaction models can be edited in the core operations module. Adjustments in the reaction rate constants are done in the calibration module. Models can be simulated in two ways: (1) through the Netlogo setup and go procedure buttons and (2) using the built-in behavior space tool. In both cases, output time evolution data on all the agent species may be saved in text/csv CSV format. Octave/ MATLAB functions are in the support scripts module to provide averaging between runs as well as output consolidation and visualization. That gets into the reactor to the attenuated levels due to shadowing. This is implemented by programmatically reducing the amount of incident photon agents from the light source in relation to an increase in the catalyst agents in simulation.

6.3.1.1 Regression of Hydrogen Production by Irradiance and Catalyst Concentration

Agent based simulations such as PHyRe treat the entities of a model as computer agents, where agents are defined as programmed autonomous members of a virtual environment. For this model, an agent may be any one of the following entities, a photon, an aqueous molecule, a catalyst active site or a hydrogen gas molecule.

A key component of the model is the way it adapts the number of photons representing computer agents to the initial amount of catalyst computer agents in the reactor simulation environment. Physically, the fewer solids there are dispersed in the solution, the more light is accessible to the various parts of the reactor. In the model, the number of photon computer agents is made to be a function of the number of catalyst computer agents, and this number is recomputed at every timestep.

To map out a relationship between the light getting inside the reactor and the amount of catalyst being stirred inside, we used experimental data in Section A.2.3, where hydrogen production was measured at varying initial catalyst amounts. We compared that with simulation output of hydrogen for combinations of initial populations of photon computer agents and catalyst computer agents. The output of the simulator was used to create a least squares regression equation mapping the experimental hydrogen produced to the simulators' initial-photon computer agent population and to the initial catalyst computer agent population.

The NetLogo variables used in the program are named: initial-photons, initial-CdZnS, initial-SO_3 and initial-S. These correspond to concentrations of photons, catalyst, sulfite and sulfide, respectively. The Netlogo behavior space values for these variables used for the simulation set are shown in Algorithm 6.1. A simulation was run for every combination of values listed. The results can be found in tabular form in Section A.3. Due to the larger variation in values of initial-photons, we included a natural logarithmic predictor term for every value relating to this variable. We also included an interaction term for initial-photons and initial-CdZnS in the regression predictors.

Algorithm 6.1: Behavior space variables set

```
["initial-photons" 800 490 360 240 140 60 28 7 4 14]
["initial-CdZnS" 12000 9000 8000 7000 6000 5000 4000 2000]
["initial-S03" 7000]
["initial-S" 12000]
```

The form of the regression equation used for the model is 16. The different k's stand for the constant coefficients that regression analysis will solve for. The initial-photon data are represented by hv. The initial-catalyst data are represented by cat. The interaction term is $k_3 ln(hv)ln(cat)$. H_x represents the evolved hydrogen by the fourth hour. The computed values for the coefficients k0–k6 are in Table 6.1. The numbers may differ so much in scale that they become more liable to round-o and truncation errors. This can be avoided by redefining cat^2 to be $cat2 = (cat^2/10000)$. The resulting t-test is given by the regression statistics in Table 6.2.

$$H_2 = k_0 + k_1 ln(hv) + k_2 ln(cat) + k_3 ln(hv) \times ln(cat) + k_4 ln(cat)^2 + k_5 cat + k_6 cat^2$$

$$(6.16)$$

TABLE 6.1
Regression Coefficients

$k0$	Intercept	15384.539155322434
$k1$	lnhv	−529.4492745872622
$k2$	lncat	−3744.914493920918
$k3$	lnhv-lncat	64.00729638658983
$k4$	lncat^2	235.63705474921838
$k5$	initial-CdZnS	−0.16258615906749382
$k6$	cat2	0.07812315227316402

TABLE 6.2
Regression Statistics

Multiple R	0.9858317160242239
R^2	0.971864172319266
Standard Error	29.46757015295423
Adjusted R^2	0.9694175786078979
Observations	76

The effective amount of light in the system due to shadowing effects can be solved using the equation relating hydrogen production, initial-photons and initial-catalysts. With some algebraic manipulation, Eq. 6.16 becomes Eq. 6.17.

$$hv = exp \frac{H_2 - k_0 - k_2 ln(cat) - k_4 ln(cat)^2 - k_5 cat - k_6 cat2^!}{k_1 + k_3 ln(cat)}$$ (6.17)

6.3.1.2 Catalyst to Photon Line Fitting

To use Eq. 6.17, the physical units of the experimental data were converted to the equivalent number of objects in the agent-based model. For instance, 0.1 g of CdZnS catalyst can be 6000 catalyst objects, 0.1 M sulfide can be 12000 S objects and 0.1 M sulfites can be 7000 SO_3 objects. Then run a simulation with an input of 6000 catalyst objects, 12000 S objects and 7000 SO_3 objects. The number of hydrogen objects produced at the timestep equivalent to 4 hours will correspond to the amount of hydrogen already produced at the fourth hour of the experiment. For instance, if in an experiment we produced 9.76 micromoles of hydrogen at hour 4 and simulation outputs 80 H_2 objects at the equivalent timestep, it can be said that the conversion rate is 9.87 micromoles of hydrogen gas for every 80 H_2 objects simulated.

Experiments were conducted that relate the hydrogen evolved at the fourth hour with varying initial CdZnS amounts. These physical values were converted to equivalent numbers of model objects in Table 6.3. The number of effective hv expected to be in the model to produce the equivalent experimental hydrogen gas was computed from Eq. 6.17.

Upon generation of enough data points, regression was made to mathematically model the relationship of the amount of catalyst in the experiment to the amount of effective light inside the reactor with respect to the shadowing caused by the solid catalyst particles. The trendline and equation for the plausible mathematical are shown in Figure 6.7.

Light in the agent-based simulation is modeled as objects which are continuously made and move across the simulation environment. Increasing its number, especially in power law fashion, can cause the computational model to slow or stop. The simulation hardware used was tested to have a limit of 1000 photon agents. The corresponding effective catalyst agent range for the shadowing mathematical model is from $5000 < cat < 12,000$ agents, as can be drawn from Figure 6.7. This range is used to check whether an experiment is in the range that can be simulated by our computational model.

TABLE 6.3
Relation of the Amount of Catalyst to the Effective Light in the Model

H_2 (μmol)	CdZnS (g)	H_2 objects	Catalysts	hv
2.0382	0.075	67	4500	11,763
2.4409	0.1	80	6000	140
2.4335	0.2	80	12,000	1

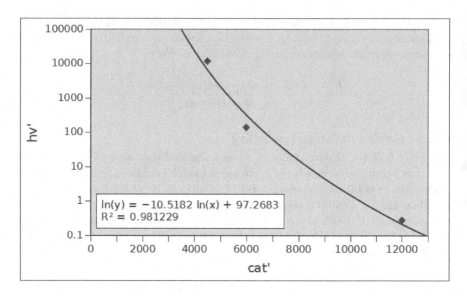

FIGURE 6.7 Power law regression trendline for determining initial-photons from initial-catalyst values.

The equation relating photons (hv) with catalysts (cat) forms the trendline shown in Figure 6.7. This equation is created through the least-squares setting and was programmed into NetLogo.

6.3.2 COMPENSATIONS

6.3.2.1 Rate Law Compensation

PHyRe 1.3 introduces the ability to define boundaries and constraints that modify the effects of the rate laws. These functions compensate for the physical phenomena not captured by the reaction mechanism. Compensation functions are integrated in the NetLogo Code. Functions are defined using observables, which compute the concentrations of species with specific properties [13]. The functions are evaluated either globally, over the entire system, or locally, over a specific reagent or chemical species. Local functions greatly expand the range of agreement between the simulation and the experimental data, because they enable a single rule to affect the rate of the reactions with rates that depend on the specific properties of the reacting species [14].

6.3.2.2 Accelerated Simulation through Population Reduction

The stochastic simulation of agent-based models slows as the number of computer agents representing chemical species increases. The simulation becomes computationally expensive, since computer memory usage increases linearly with the number of programmed agents [15]. However, increasing agent population sizes bring the individual-based model closer to the molecular system level that the model seeks to describe. PHyRe 1.3 has maintained accuracy while significantly increasing simulation speed from upwards of one hour to below 10 minutes for an average simulation compared to the previous version. This was accomplished by reducing the population of all the species and making the simulation environment proportionally smaller while maintaining the ratio of the different agent entities. This resulted in an approximate five-fold increase in simulation speed. The emergent behavior of the different species was tested to have maintained the level of agreement with experimental results as before the proportional population reduction.

6.3.2.3 Hybrid Particle/Population (HPP) Simulation for Electrons and Holes

Network-free simulation methods which do not enumerate species and reactions are advantageous in simulating complex models [7, 8]. The limitation of this technique is that memory usage increases linearly with the number of particles. Hogg et al in 2014 showed that treating a set of species as population variables rather than individual particles avoids the memory costs associated with having pools of identical particles. This hybrid method called hybrid particle/population (HPP) was shown to significantly reduce computational memory expense with no significant negative effects on simulation accuracy and run time [15]. A similar technique is used in PHyRe 1.3 with the special objective to reduce runtime while maintaining memory and accuracy.

During tests to extend the range of reagent concentrations for which the computational model is useful, it was discovered that when the reactor is in a sulfide-starved state, meaning there is a low concentration of the hole scavengers in the reactor, recombination reaction of electrons and holes happens so fast that in order to account for this large relative speed difference, it was determined that the reactions of other particles have to be slowed down many times their current speeds. This slows the entire simulation process. HPP was used to remove the need to induce this slowdown. Accuracy was increased while the simulation speed was maintained.

6.3.3 ADDITIONAL SUPPORT SCRIPTS AND FEATURES

PHyRe 1.3 also has a number of additional features to those described above. It contains scripts that average the repeats of an experiment as input to a Markovchain Monte-Carlo probabilistic sub-model that infers optimum species concentrations based on probabilistic graphical methods [16]. Species concentration visualization scripts also exist in order to compare the time evolution of up to three species against each other, shown as two-dimensional or three-dimensional graphs. A rate script also exists to visualize the rate of hydrogen production with respect to the concentration of any reactant species at a particular time in the experiment. Conversion scripts were created to provide a mapping between any of the reagent or product species' physical micromole units to computer agents, and vice versa. Long simulations can now also be terminated automatically when the probability of further hydrogen production is computed to be below a threshold value. Parameter estimates for the reaction rates, initial reagent concentrations and catalyst amounts can now also be saved as initialization sets that allow efficient re-simulation of chosen past experiments.

6.4   RESULTS

The resulting graphs are shown in Figure 6.8. Experiment set A consists of a single experiment where hydrogen measurements were done every hour for twelve hours. The simulation was able to capture the form of the power law form of the graph. The average relative error is 22.84% for the chosen simulation parameter range of two to twelve hours. Experiment set B consists of six batch runs of varying initial sulfide concentrations. The simulation output was able to capture the general trend of the graph. The average relative error is 31.17% for the chosen simulation parameter range of 0.005 to 0.1 molars of initial [S^{2-}]. Experiment set C consists of five batch runs of varying sulfite initial concentrations. As seen in Figure 6.8, the simulation output was able to accurately track the experimental data. The average relative error is 10.67% for the simulation parameter range of 0.01 to 0.3 molars of initial sulfite. Experiment set D contains four experiments with varying catalyst amounts. The simulation output follows the trend of the experimental data. The average relative error is 19.16% for the chosen simulation parameter range of 0.075 to 0.2 grams of CdZnS powdered catalyst. All in all, sixteen batch run experiments were used to calibrate the simulator.

The occurrence of the reactions is stochastically determined using a probability function, which depends on different parameters computed via random number generators. Each run, therefore, produces a different sequence of probabilistic events and slightly different hydrogen production values, a characteristic agent-based models share with actual laboratory experiments. At every timestep, a normal distribution with mean corresponding to the expected probability that a reaction will occur given collision of reactants is sampled as a condition for executing

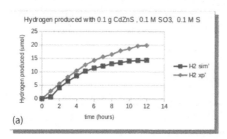
(a)

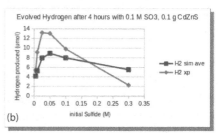
(b)

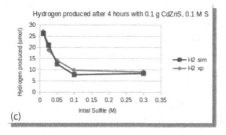
(c)

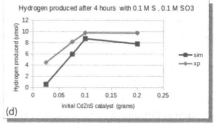
(d)

FIGURE 6.8   Recalibration experiments.

the reaction procedure. The collision of reactants, however, is determined by the autonomous computer agent movement inside the virtual reactor, and this makes it different from the stochastic simulation algorithm or Gillespie's method. This is implemented in NetLogo. Repeat simulations are done, and the average values were used in this report.

It has been verified that the model can recreate the trends and trajectories of the hydrogen production graphs with respect to all the parameters of interest. As an optimizing step, 182 simulations were run to search for optimal starting concentrations of reactants in order to maximize hydrogen production for a nine-hour batch run. The same reactor setup and configuration were used in the published [4] baseline experiment to validate the simulated optimized experiment. The independent variables were the sulfide and sulfite initial concentrations. All the other parameters were kept the same, as were the preparation of the catalyst and the solution. Each circle in Figure 6.9 represents a simulation. The color of the circle provides an estimate of the expected hydrogen gas produced at the ninth hour. Redder circles correspond to higher hydrogen production. Bluer circles correspond to lower hydrogen production. More simulations were done where higher hydrogen production was expected and where there are experimental data available. The relative differences in hydrogen production with respect to the initial reactant concentrations are shown in Figure 6.10. The independent axes are in molars while the hydrogen is in micromoles of gas.

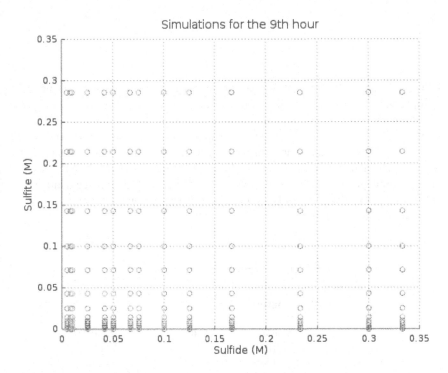

FIGURE 6.9 Searching for optimal initial concentrations.

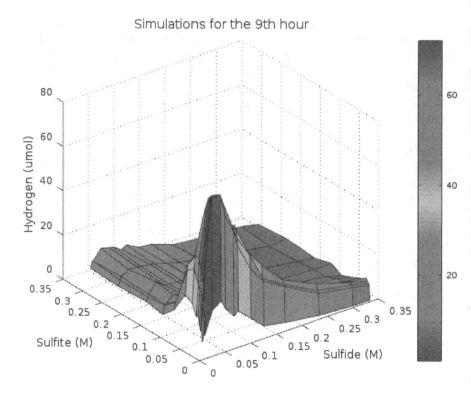

FIGURE 6.10   Surface map of simulation data.

The relevant parameters of the baseline experiment vis à vis the parameters of the optimized experiment are shown in Table 6.4. All parameters are within model-tested ranges.

The results are shown in Figure 6.11. Graphs C and D are simulation and experimental plots from the baseline data set used to calibrate the model. Graphs A and B are the optimal simulation results and the validation experiment results. It shows that the simulation tracks the experimental results well with an average relative error of less than 15%. The results also indicate a hydrogen production increase of 300% from baseline at the ninth hour.

TABLE 6.4

Experiment Parameters

	Catalyst (g)	Sulfide (M)	Sulfite (M)
Baseline	0.100	0.100	0.100
Optimized	0.100	0.017	0.010

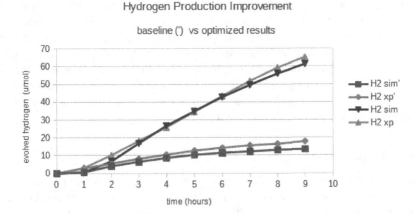

FIGURE 6.11   Results of optimization.

6.5   DISCUSSION

In this section, we discuss the results of using ABM to model, simulate, optimize and design a photocatalytic hydrogen production experiment to produce more than a 300% increase in output compared to baseline, while using fewer reagents and thus less cost. There arefew mathematical models like those in [4, 5] and, to the best of our knowledge, no computational models of photocatalytic hydrogen production have been developed to date, and none, that we are aware of, study directly catalyst poisoning by sacrificial reagents leading to decreased hydrogen production capability.

The obtained results confirmed that the initial concentrations of sacrificial reagents can significantly affect hydrogen production rates. Sacrificial reagents react with chemical species that would have competed with hydrogen production processes if left uncontrolled. Sulfide is a hole scavenger. Its presence promotes reactions that remove holes. This minimizes electron-hole recombination, freeing the electrons to further hydrolysis reactions. Sulfites react with the colored polysulfide species in order to produce transparent thiosulfates and hydrosulfides. These products will not compete with the photon absorption reaction, unlike the colored polysulfide species. Though the sacrificial reagents perform a critical function in producing hydrogen, they also cause catalyst active site poisoning. An optimal control of the reagent concentrations is not simple, because many reactions directly or indirectly modify reagent concentration in the course of a batch run. The model we reported contributes to clarify the influence of each of the reagents on the amount of hydrogen produced. Moreover, the power of this computational model permits predicting the effects of changing multiple reagents at the same time. The use of computational models will, in the future, allow engineers to monitor the dynamics of photocatalytic hydrogen production and its response to perturbations and interventions. The changes in the absorbance in the reactor due to different species absorbing different wavelengths were indirectly inferred through the resulting changes in hydrogen gas production.

The model was tuned with the experimental data collected by Tambago et al where hydrogen production with different values of reagents were analyzed [4]. In such experimental conditions, PHyRe 1.3 reproduces the same graph trends and plot tendencies as in the laboratory data. In the present model, most of the bulk effects and outcomes are driven by the programming of the micromechanics as entity behavior. We statistically modeled the light-scattering effects of high catalyst concentration. We recognize the importance of this effect, and the PHyRe simulator is extendable to include a rule-based light scattering model should data for it become available. Future updates of the model may take the roles of light scattering and powdered catalyst shadowing into account as agent rules.

6.6 CONCLUSION

We presented a model that describes the role of sacrificial reagents in photocatalytic hydrogen production. In silico experiments provided optimal reagent concentration values that produced a more than 300% increase in hydrogen production when validated in the laboratory. The model and its computer implementation are flexible, and new chemical entities and interactions can easily be added to the model.

Moreover, the model produced an important suggestion for future laboratory experiments on the effects of sacrificial reagents to hydrogen production. High initial concentrations of reagents are detrimental to the desired hydrolysis reaction. In this condition, the model should be useful not only in predicting optimal initial reagent concentrations, but also in predicting durations by which these sacrificial reagents will eventually be consumed. As PHyRe simulations predictions are experimentally validated, it will be possible to obtain important information on the dynamics of the reactions, thereby getting insight in how to better to optimize hydrogen production. Results will be published in due course.

PHyRe 1.3 is an active software model project with ongoing development efforts that include porting the model to a three-dimensional, agent-based simulation environment. Future work includes improving support for software modeling community standards such as the standard for annotation of rule-based models [17]. Future development also includes a parameter estimation tool for models simulated by the PHyRe software.

6.7 FURTHER WORK

Work is currently being undertaken to translate the model into three dimensions to see how much more accurately it can describe the system through the additional dimension with the extra complexity and computational effort involved.

Recalibration of the model with respect to experiments using monodispersed catalysts will provide significant improvement in the modeling of light scattering effects. The monodispersed colloid will produce a much better characterized system that may have catalyst light attenuation that conforms to the Beer-Lambert law. Should the monodispersed colloids be less than 1 micron in diameter, light scattering is expected to be negligible, and the catalyst can be supported in solution with mechanical stirring.

Fluorescent and absorption spectroscopy may be used to track the changes in the concentrations of the intermediates and improve the accuracy of the model.

Spectroscopy may also characterize how the different species absorb different wavelengths of light and result in absorbance changes in the reactor.

The modeling can be extended to oxygen as well, using oxygen membrane polarographic electrodes to measure oxygen evolution in real time.

ACKNOWLEDGMENTS

The team would like to acknowledge the financial and motivational support of the Engineering Research and Development for Technology (ERDT) and the Chemical Engineering Foundation. We are grateful to the Scientific Computing Laboratory (SCL) of Computer Science for the computing equipment and to the Fuels, Energy and Thermal Systems (FETS) Laboratory of Chemical Engineering for housing the reactor and hosting all the laboratory experiments.

RVLC wishes to deeply thank Dr. Kevin L. Henbest of the Department of Chemistry of the University of Oxford and Dr. Prospero Naval of the Department of Computer Science of the University of the Philippines for their keen observations and inputs to this research.

VPB specially thanks Dr. Kevin L. Henbest of the Department of Chemistry of the University of Oxford for the long hours of useful scientific discussions.

REFERENCES

1. M. Ni, M. K. Leung, D. Y. Leung, and K. Sumathy, A review and recent developments in photocatalytic water-splitting using TiO_2 for hydrogen production, Renewable and Sustainable Energy Reviews, vol. 11, pp. 401–425, 2007.
2. J. Yu, Y. Hai, and M. Jaroniec, Photocatalytic hydrogen production over CuO-modified titania, Journal of Colloid and Interface Science, vol. 357, pp. 223–228, 2011.
3. O. Rosseler, M. V. Shankar, M. K.-L. Du, L. Schmidlin, N. Keller, and V. Keller, Solar light photocatalytic hydrogen production from water over Pt and Au/TiO_2(anatase/rutile) photocatalysts: Influence of noble metal and porogen promotion, Journal of Catalysis, vol. 269, no. 1, pp. 179–190, 2010.
4. H. M. G. Tambago and R. L. D. Leon, Intrinsic kinetic modeling of hydrogen production by photocatalytic water splitting using cadmium zinc sulfide catalyst, International Journal of Chemical Engineering and Applications, vol. 6, pp. 220–227, 2015.
5. J. Sabate, S. Cervera-March, R. Simarro, and J. Gimenez, Photocatalytic production of hydrogen from sulfide and sulfite waste streams: A kinetic model for reactions occurring in illuminated suspensions of CdS, Chemical Engineering Science, vol. 45, no. 10, pp. 3089–3096, 1990.
6. F. Pappalardo, P. Zhang, M. Halling-Brown, K. Basford, A. Scalia, A. Shepherd, D. Moss, S. Motta, and V. Brusic, Computational simulations of the immune system for personalized medicine: State of the art and challenges, Current Pharmacogenomics and Personalized Medicine, vol. 6, pp. 260–271, 2008.
7. L. A. Chylek, L. A. Harris, C.-S. Tung, J. R. Faeder, C. F. Lopez, and W. S. Hlavacek, Rule-based modeling: A computational approach for studying biomolecular site dynamics in cell signaling systems, Wiley Interdisciplinary Reviews: Systems Biology and Medicine, vol. 6, pp. 13–36, 2014.
8. L. A. Chylek, L. A. Harris, J. R. Faeder, and W. S. Hlavacek, Modeling for (physical) biologists: An introduction to the rule-based approach, Physical Biology, vol. 12, p. 045007, 2015.

9. U. Wilensky, Center for Connected Learning and Computer-Based Modeling. Northwestern University, Evanston. NetLogo, 1999.

10. R. V. L. Canseco, V. P. Bongolan, H. M. G. Tambago, and R. L. D. Leon, Agent-Based Modeling of Visible Light-driven Photocatalytic Hydrogen Production using CdS, in 19th Regional Symposium of Chemical Engineering, Bali, Indonesia, pp. 1–5, 2012.

11. R. V. L. Canseco, V. P. Bongolan, H. M. G. Tambago, and R. L. D. Leon, Calibration of a Rule-Based Computational Model of Photocatalytic Hydrogen Production using Cadmium Sulfide Catalyst with Sulfide and Sulfite Reagents, in 8th National Workshop on Modeling, Simulation and Scientific Computing (MODEL 2015), Quezon City, Philippines, pp. 1–4, Special Interest Group in Modeling, Simulation and Scientific Computing (SIG-MODEL Philippines), 2015.

12. K. R. Tolod, C. J. E. Bajamundi, R. L. de Leon, P. Sreearunothai, R. Khunphonoi, and N. Grisdanurak, Visible light-driven photocatalytic hydrogen production using Cu-doped $SrTiO_3$, energy sources, Part A: Recovery, Utilization, and Environmental Effects, vol. 38, no. 2, pp. 286–294, 2016.

13. J. R. Faeder, M. L. Blinov, and W. S. Hlavacek, Rule-based modeling of biochemical systems with BioNetGen, Methods in Molecular Biology, vol. 500, no. 1, pp. 113–167, 2009.

14. M. W. Sneddon, J. R. Faeder, and T. Emonet, Efficient modeling, simulation and coarse-graining of biological complexity with NFsim, Nature Methods, vol. 8, pp. 177–183, 2011.

15. J. S. Hogg, L. A. Harris, L. J. Stover, N. S. Nair, and J. R. Faeder, Exact hybrid Particle/Population simulation of rule-based models of biochemical systems, PLoS Computational Biology, vol. 10, p. e1003544, 2014.

16. R. V. L. Canseco, and V. P. Bongolan, Markov Chain Monte Carlo optimization of visible light-driven hydrogen production, in TENCON 2012 IEEE Region 10 Conference, pp. 1–5, IEEE, 2012.

17. G. Misirli, M. Cavaliere, W. Waites, M. Pocock, C. Madsen, O. Gilfellon, R. Honorato-Zimmer, P. Zuliani, V. Danos, and A. Wipat, Annotation of rule-based models with formal semantics to enable creation, analysis, reuse and visualization, Bioinformatics, vol. 32, no. 6, pp. 908–917, 2015.

APPENDIX

A.1 REACTOR SETUP

We look at the physical reactor setup in order to determine the initial geometries to defi ne the agent-based simulation environment.

The schematic diagram for the experimental setup is shown in Figure 6.12. Photocatalytic reactions are carried out in a Pyrex beaker-like vessel with an O-ring joint seal. The inner diameter is 5 cm. With 100 mL of catalyst suspension, the height of the liquid in the glass is about 5 cm. The headspace has a volume of 126 cm^3. At the top of the headspace is the sampling port with a polytetra uoroethelyn (PTFE) septum. The lamp is a 400-W halogen lamp (Osram Holaline Eco SST) with an ultraviolet light filter (Kenko MC UV, >390 nm), allowing only visible light to reach the reactor. The lamp-facing side of the reactor has a ground glass texture in order to mimic diffuse incoming radiation. The magnetic stirring is set to 360 rotations per minute. A picture of the reactor running a batch-mode experiment is shown in Figure 6.13.

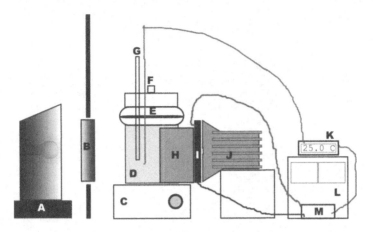

FIGURE 6.12   Reactor schematic. (A) lamp, (B) UV cut-o lter, (C) magnetic stirrer, (D) Pyrex reactor, (E) reactor headspace, (F) gas sampling port, (G) port For N_2 purging, (H) cold sink, (I) Peltier cell, (J) heat sink, (K) temperature controller, (L) power supply, (M) mechanical relay.

Lamp irradiance was measured using a DayStar DS 05 solar meter. Lamp spectral irradiance was determined by an Ocean Optics HR2000+ high-resolution spectrometer. Reaction temperature was maintained at 25 0.5°C using a Peltier cell cooling system. The solution inside the reactor was composed of $Cd_{0.4}Zn_{0.6}S$ photocatalyst particles continuously stirred with aqueous sodium sulfide (Na_2S) and sodium sulfite (Na_2SO_3), in 100 mL of deionized water.

FIGURE 6.13   Reactor picture.

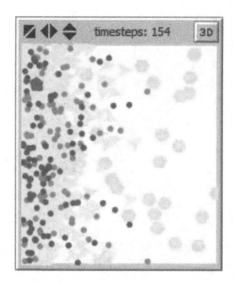

FIGURE 6.14   Reactor environment in Netlogo.

A cross section of the Pyrex vessel of the reactor can be modeled by a two-dimensional agent environment. Colored shapes can represent particles as shown in Figure 6.14. The yellow hexagons represent catalyst active sites. The right-facing triangles represent light particles moving from a light source located to the left of the environment. Orange dots represent electrons. Blue particles represent holes. Both electrons and holes initially cluster to the left of the environment as they are created through the collision of the randomly located catalysts and left-entering light. The snapshot shows the early stage of a simulation at timestep 154. The configuration of the agent environment should generally follow the configuration of the physical setup.

A.2   EXPERIMENTAL DATA USED

A.2.1   Hourly Hydrogen Production Data

Hour	umol H_2
1	2.8217
2	5.4950
3	8.1188
4	10.3960
5	12.6237
6	14.2574
7	15.5940
8	16.4851
9	17.8217
10	18.5643
11	19.6039
12	19.8019

A.2.2 Sulfite Experiment Data

SO_3 M	H_2 umol
0.0097472924	6.7457
0.0245487365	4.7033
0.0494584838	3.5254
0.0992779783	2.4322
0.2996389892	2.2542

A.2.3 Catalyst Experiment Data

CdZnS	H_2 umol
0.025	1.1108
0.075	2.0381
0.1	2.4408
0.2	2.4334

A.2.4 Sulfide Experiment Data

S	H2
0.005	1.3436
0.01	2.2796
0.025	3.2926
0.05	3.2571
0.1	2.4514
0.3	0.5675

A.3 DATA FOR REGRESSION ANALYSIS

This solves for $H_{2,final} = F(hv_i, cat_i)$. All the combinations that resulted in 0 hydrogen objects at the 4th hour were removed from the table. They were not used in the calculations because they distort results.

Run #	initial-photons	initial-CdZnS	H_2	Run #	initial-photons	initial-CdZnS	H_2
1	800	12000	597	37	140	4000	16
2	800	9000	302	38	140	2000	3
3	800	8000	208	39	60	12000	514
4	800	7000	167	40	60	9000	248
5	800	6000	109	41	60	8000	182
6	800	5000	47	42	60	7000	115
7	800	4000	33	43	60	6000	50
8	800	2000	4	44	60	5000	29
9	490	12000	598	45	60	4000	4
10	490	9000	279	46	28	12000	503
11	490	8000	203	47	28	9000	198

(Continued)

Run #	initial-photons	initial-CdZnS	H_2	Run #	initial-photons	initial-CdZnS	H_2
12	490	7000	155	48	28	8000	111
13	490	6000	87	49	28	7000	69
14	490	5000	57	50	28	6000	30
15	490	4000	27	51	28	5000	15
16	490	2000	4	52	28	4000	7
17	360	12000	621	53	7	12000	237
18	360	9000	295	54	7	9000	92
19	360	8000	251	55	7	8000	44
20	360	7000	147	56	7	7000	32
21	360	6000	80	57	7	6000	17
22	360	5000	43	58	7	5000	2
23	360	4000	22	59	7	4000	4
24	240	12000	622	60	4	12000	148
25	240	9000	274	61	4	9000	53
26	240	8000	206	62	4	8000	24
27	240	7000	159	63	4	7000	18
28	240	6000	80	64	4	6000	2
29	240	5000	47	65	4	5000	2
30	240	4000	23	66	14	12000	340
31	140	12000	589	67	14	9000	153
32	140	9000	298	68	14	8000	81
33	140	8000	227	69	14	7000	46
34	140	7000	149	70	14	6000	35
35	140	6000	86	71	14	5000	14
36	140	5000	48				

7 Mathematical Modelling on Non-Ideal Behavior of Different Bioreactor Configurations

Santanu Sarkar

7.1 INTRODUCTION

Biological processes are adopted by almost all industries starting from synthesis of a wide range of products to wastewater treatment. Hence, bioreactors have wide applications in different sectors like pharmaceuticals, food, beverage, bio-refineries and wastewater treatment. It is obvious that the intrinsic parameters, such as dilution rate, inlet substrate concentration, growth rate of biomass, biomass concentration, etc., influence product yield from any biological processes and those parameters control the performance of a bioreactor. The variations of those parameters should be considered during the design of a reactor to achieve maximum performance from it. Biochemical systems are more complex and slowly processed in nature compared to normal ones with higher residence time. Moreover, the characteristics of the reaction medium vary with variation of living species that take part during the reaction. Several types of bioreactors are used commercially, depending on applications and end products. However, bioreactors are mainly classified in three categories based on the occurrence of the reactions: continuous stirred tank bioreactors (CSTBRs), laminar or plug flow bioreactors (PFBRs) and packed bed bioreactors (PBBRs). There are literature and models available to describe non-ideality of common chemical reactions in different types of reactors. However, only a few studied have reported on measurement of hydrodynamic and non-ideal parameters for biochemical system, as such processes are more complex than normal chemical ones. More often, the viscosity of the reaction medium gradually increases with growth of biomass; thus, non-ideality of bioreactors becomes more prominent [1–10]. Moreover, non-ideality creeps in mainly due to imperfect mixing of the whole reaction mass, leading to a fraction of dead volume in the reactor. Sometimes, due to improper placement of inlet and outlet nozzles, uneven packing and formation of biofilms, a portion of feed may just bypass the reactor without undergoing any reaction [5–8]. Therefore, it is necessary to measure and incorporate non-ideal parameters for designing biological reacting system, considering the change of hydrodynamic behavior due to growth of biomass [5–15].

Very little literature is available on no-ideality and residence time distribution (RTD) on bioreactors. Considering growing applications of bioreactors in different

DOI: 10.1201/9781003435181-7

areas in commercial scales, it is important to discuss such pitfalls of bioreactors. This chapter focuses on the basics of non-ideality and modellings to measure the non-ideal parameters. It also deals with actual hydrodynamic behaviors of real bioreactors.

7.2 BASIC OF NON-IDEALITY

Any kind of reactor has been modelled as an ideal reactor. However, in the real scenario, it has been observed that its behavior is very different from the expected value. Some factors make up the flow pattern, such as the residence time distribution (RTD) of the culture medium, the state of aggregation of the flowing material, its tendency to clump and for a group of molecules to move about together and the nature of the mixing of material inside the reactor. The deviation of a system from the ideality happens due to channeling, bypassing and formation of the dead zone [5–17].

7.2.1 THE RESIDENCE TIME DISTRIBUTION

The amount of time spent by a reacting molecule inside the reactor is called residence time for that reactant molecule [16, 17]. The formation of a dead zone and channeling inside the reactor affects the residence time of the reacting materials. In the dead zone, the rate of reaction is very slow, or it is stopped. Due to channeling, some reactants just bypass the reacting or perfectly mixed zone. So it is quite normal that elements of reactants taking different routes through the reactor may take different lengths of time to pass through the vessel.

7.2.1.1 Exit Age Distribution Function E(t) of RTD

The distribution of the time spent by each molecule inside the reactor, called exit age distribution E(t), has a unit of time^{-1} [16, 17]. It is obvious that the following Eq. (7.1),

$$\int_0^\infty E(t)\,dt = 1 \tag{7.1}$$

7.2.1.2 Cumulative Distribution Function F(t) of RTD

The tracer concentrations were measured using a spectrophotometer. The cumulative distribution function F(t) can be evaluated using measured data from the step experiments. The F(t) values were calculated using Eq. (7.2) [16, 17].

$$\left[\frac{C_{out}}{C_0}\right]_{step} = \int_0^t E(t')\,dt' = F(t) \tag{7.2}$$

7.2.2 MEASUREMENT OF RTD

RTD can be determined experimentally by injecting an inert chemical, molecule or atom, called a tracer, into the reactor at some time (t = 0) and then measuring the tracer concentration at the exit stream as a function of time. Pulse and step inputs are the two commonly used methods of injection [16, 17].

7.2.2.1 Pulse Experiment

In a pulse input, an amount of tracer is suddenly injected in one shot into the feed stream, entering the reactor in as short a time as possible. The outlet concentration is then measured as a function of time. The effluent concentration-time curve is referred to as the C curve in the RTD analysis. E(t) describes in a quantitative manner how much time different fluid elements have spent in the reactor. The volumetric flow rate at steady state is constant, and E(t) can be defined as using Eq. (7.3).

$$E(t) = \frac{C(t)}{\int_0^\infty C(t)\,dt} \tag{7.3}$$

7.2.2.2 Step Experiment

The tracer is fed to the system with constant volumetric flow rate maintaining the same tracer concentration to analyze a step input to measure the cumulative distribution F(t) [16, 17]. Consider a constant rate of tracer addition to a feed that is initiated at time $t = 0$. The output concentration from a vessel is related to the input concentration by the convolution integral form of Eq. (7.4).

$$C_{out} = \int_0^t C_{in}(t - t')E(t')\,dt' \tag{7.4}$$

As $C_{in} = C_0$, the inlet tracer concentration is constant so, Eq. (7.4) can be written as

$$\left[\frac{C_{out}}{C_0}\right]_{step} = \int_0^t E(t')\,dt' = F(t) \tag{7.5}$$

7.2.3 Mean Residence Time

The mean residence time is described as the ratio of volume (V) of the reactor to the volumetric flow rate (v) [16, 17]. Then it is also called space time or average residence time ($\tau = V/v$). Considering the RTD for a non-ideal reactor, the mean residence time is defined as follows Eq. (7.6),

$$t_m = \frac{\int_0^\infty tE(t)\,dt}{\int_0^\infty E(t)\,dt} = \int_0^\infty tE(t)\,dt \tag{7.6}$$

7.2.4 Other Moments of RTD

Except mean residence time there is another variance. The magnitude of variance is the measure of spread of distribution; the greater the value of this moment is, the greater a distribution's spread will be. It can be defined as using Eq. (7.7).

$$\sigma^2 = \int_0^\infty (t - t_m)^2 \, E(t) \, dt \tag{7.7}$$

7.2.5 Different Mathematical Models for Bioreactor Designing

7.2.5.1 Dead-Space-and-Bypass Model (Two-Parameter Model)

A real bioreactor might be modeled by different combination ideal reactors. There are two zones in any reactor; one is perfectly mixed, and another portion is a dead zone. Moreover, some portions of feed leave the without taking part in the reaction, called bypassing [16, 17]. Model parameters could be obtained from step experiments. By simple mass balance, the following derivation can be obtained using Figure 7.1.

$$V_s \frac{dC_{TS}}{dt} = v_s \left(C_{T0} - C_{TS} \right) \tag{7.8}$$

α = the fraction of total volume of the reactor, which actually takes part in the reaction; β = the fraction of the feed flow rate, which is bypassed; V = the volume of the CSTBR; V_s = the effective volume of the CSTBR = αV; V_d = dead volume = $(1 - \alpha)$ V; C_{T0} = input tracer concentration; C_{TS} = output tracer concentration from the reactor; C_T = tracer concentration at effluent; v_0 = tracer flow rate at inlet; v_s = actual tracer flow rate to the reactor = $(1 - \beta)$ v_0; v_b = tracer bypassing flow rate = βv_0; τ = residence time = V/v_0.

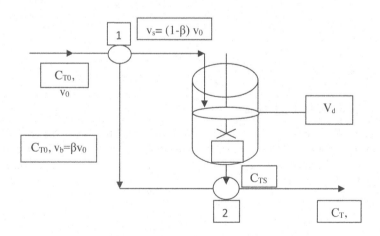

FIGURE 7.1   Schematic diagram for the two-parameter model system.

The conditions for the positive step input are at $t < 0$, $C_T = 0$ and at $t \geq 0$, $C_T = C_{T0}$. The balance around the junction point (2) results Eq. (7.8) and (7.9).

$$C_T = \frac{v_b C_{T0} + C_{TS} v_s}{v_0} \tag{7.9}$$

On integration of Eq. (7.7) and substituting in terms of α and β gives

$$\frac{C_{TS}}{C_{T0}} = 1 - \exp\left[-\frac{1-\beta}{\alpha}\left(\frac{t}{\tau}\right) \right] \tag{7.10}$$

Combination of Eqs. (7.9) and (7.10) gives,

$$\frac{C_T}{C_{T0}} = 1 - (1-\beta)\exp\left[-\frac{1-\beta}{\alpha}\left(\frac{t}{\tau}\right) \right] \tag{7.11}$$

On rearranging and taking logarithm, Eq. (7.11) reduces to Eq. 7.12,

$$\log\frac{C_{T0}}{C_{T0} - C_T} = \log\frac{1}{1-\beta} + \frac{1}{2.303}\left(\frac{1-\beta}{\alpha}\right)\frac{t}{\tau} \tag{7.12}$$

A real CSTB can be modeled using the above-mentioned two-parameter model (bypassing and dead space) [6] and a developed model can be helpful for designing a real CSTBR.

7.2.5.2 The Dispersion Model

The dispersion model helps to understand axial dispersion or longitudinal dispersion and molecular diffusion inside the reactor. It can be measured with the help of pulse RTD experiments [16, 17]. Moreover, the characteristics of plug flow can be quantified with the help of the dispersion coefficient D (m^2/s). In this model, mean residence time (t_m) and variance (σ^2) are the measurable parameters.

The dispersion model has been developed with help of Fick's law, and it has been assumed that the molecules are propagating in the x-direction by all driving forces. Therefore, it can be written as,

$$\frac{\partial C}{\partial t} = D\frac{\partial^2 C}{\partial x^2} \tag{7.13}$$

where, D is the longitudinal or axial dispersion coefficient.

The dimensionless form of Eq. (7.13) can be rewritten as Eq. 7.14:

$$\frac{\partial C}{\partial \theta} = \left(\frac{D}{UL}\right)\frac{\partial^2 C}{\partial z^2} - \frac{\partial C}{\partial z} \tag{7.14}$$

where, $z = (Ut + x)/L$, and $\theta = (t/t_m) = tU/L$. D/UL represents the vessel dispersion number or inverse Peclet No (Pe). Based on the value of Pe, the extent of axial dispersion can be predicted as per below assumptions. If D/UL or Pe tends to zero or infinite, then flow is pure plug flow with negligible dispersion, and if value is the opposite, it implies the flow is mixed flow with large dispersion.

The mean residence time and variance can be calculated using Eqs. (7.15) and (7.16) in terms of tracer concentration.

$$t_m = \frac{\int_0^\infty tC\,dt}{\int_0^\infty C\,dt} = \frac{\sum t_i C_i \Delta t_i}{\sum C_i \Delta t_i} \tag{7.15}$$

$$\sigma^2 = \frac{\int_0^\infty (t - t_m)^2 C\,dt}{\int_0^\infty C\,dt} = \frac{\int_0^\infty t^2 C\,dt}{\int_0^\infty C\,dt} - t_m^2 \tag{7.16}$$

The discrete form of Eq. (7.16) can be represented as Eq. (7.17).

$$\sigma^2 \cong \frac{\sum (t - t_m)^2 C_i \Delta t_i}{\sum C_i \,\Delta t_i} = \frac{\sum t_i^2 C_i \Delta t_i}{\sum C_i \,\Delta t_i} - t_m^2 \tag{7.17}$$

A relationship was first proposed by van der Laan in 1958 [16, 17] for closed vessels as Eq. (7.18).

$$\sigma_\theta^2 = \frac{\sigma^2}{t_m^2} = 2\left(\frac{D}{UL}\right) - 2\left(\frac{D}{UL}\right)^2 \left[1 - e^{-\frac{UL}{D}}\right] \tag{7.18}$$

The above model helps to measure the value of σ^2, t_m^2 and UL/D to predict the deviation real plug flow bioreactors from ideal one.

7.3 MATHEMATICAL MODELLING OF THE CONTINUOUS STIRRED TANK BIOREACTOR

The CSTBR is a widely used bioreactor in biochemical processes, mainly in wastewater treatment and generation of useful products. The biological reaction is more complex than common chemical reactions that occur in continuous stirred tank reactors due the change of viscosity of the reaction broth [3, 6]. In 2011, Ajbar [3] investigated the non-ideality of CSTBR having recycle loops with help of static and dynamic analyses. The static model was used to find the non-linear singularity based on input and output multiplicity. Dynamic singularity helped to identify the periodic behavior of the reactor, which was investigated using Hopf points and the Jacobian matrix of the working equations. In this study, the Haldane growth rate equation was used to measure the different parameters at a steady state [3, 18, 19]. The model predicted that input multiplicity could not occur; however, output multiplicity was possible in the form of hysteresis and saddle-node for nonsterile feed and sterile feed, respectively. This model was unable to predict the periodic behavior when the yield was independent of substrate concentration but able to predict the oscillatory behavior when yield was variable. The developed model was also able to predict the effect of substrate inhibition and dilution rates on the performance of the bioreactor [3].

Bhattacharya et al. [8] had studied the non-ideality of a CSTBR using the two-parameter model (α: the fraction of total volume of the reactor, which takes part in the reaction; and β: the fraction of the feed flow rate, which is bypassed) [16, 17]. The study was conducted with the reference reaction of SO_2 degradation by *Desulfotomaculum nigrificans*. The model parameters were evaluated by conducting step and pulse experimentation in non-reacting systems. Moreover, in this study, the classical growth kinetic equation was used [8, 19, 20]. The measured value of the two-parameter model confirmed that the reactor used in the study behaved like an ideal one. Therefore, the stability analysis of the CSTBR was performed considering the reactor was ideal and the viscosity and density of the reaction medium was constant. A simple non-dimensional, non-linear mass balance equation was used for stability analysis. Local and global stability analysis was conducted by solving the Jacobian matrix and phase plane analysis respectively. The eigen values of the Jacobian matrix at certain ranges of dilution rate and substrate concentration were negative with real magnitude, which indicated a stable steady state. Moreover, the phase plane analysis also revealed that at certain dilution rates the stability of a steady state could be achieved for a certain range of substrate concentration. This study [8] has emphasized that the two-parameter model, along with stability analysis, can be very useful for designing industrial scale bioreactors. However, the study never incorporated the change of viscosity and/or concentration of the reaction medium during model development. To eliminate such a limitation, Sarkar et al. [6] conducted a study to understand the effect of change of viscosity of the reaction mixture on the non-ideality of the CSTBR.

Sarkar et al. [6] had taken the growth parameters from the growth study of *Lactobacillus sp.* in agar medium and followed the Monod model for kinetic study. In this study, a polynomic relationship has been formed between the relative viscosity and the concentration of biomass. Moreover, the step and pulse experiment was carried out at a steady state using simulated solution by varying medium viscosity. Like earlier the study [8], this study [6] also followed the two-parameter model [16, 17]. However, α and β were expressed as a function of relative viscosity (η_r). Thus, as the time progresses, the bypassing and dead volume of the bioreactor increase with the increase of biomass concentration or the viscosity of the medium. Both steady and unsteady state analysis have been performed to understand the effect of different parameters (substate concentration, dilution rate, biomass concentration, medium viscosity) on nonideality. This study can also help to understand the difference between ideal and non-ideal bio reacting systems. Mainly, the dilution rate played a major role in non-ideality, and from an unsteady state analysis, it was observed that non-ideality increased higher dilution rate. Also, steady state analysis concluded that the maximum dilution rate for a non-ideal system was significantly lower than an ideal system. The model developed in this study [6] by incorporating the medium viscosity would help to find out the values of two-parameter model those, are required for designing commercial bioreactor. However, the kinetic model and parameter can be varied depending on the species of microbes under consideration.

7.4 MATHEMATICAL MODELLING OF THE PLUG FLOW BIOREACTOR

PFBRs are used for very sophisticated purposes to produce special type of bioproducts. The performance of the PFBR widely depends on the hydrodynamic behavior of THE reaction medium. Panikov et al. [20] and Oldiges et al. [21] studied kinetics and mass transfer inside the PFBR. However, there is little literature available on hydrodynamic studies and modelling. Panikov et al. [20] had described that PFBR as much more effective than the batch system for the growth of biomass; however, they are silent on hydrodynamic behavior. As discussed earlier, the bioreacting system always deviates from the ideal system due increase in biomass concentration with time. Thus, flow pattern would deviate from ideal plug flow in PFBR. In 2005, RTD was studied for commercial PFBR used for wastewater treatment by Olivet et al. [22]. They pointed out that an RTD study might not be able to describe a mixing regime inside the reactor; however, it can be used for qualitative measurement of mixing, challenging and type of flow. Thus, the RTD study reveals hydraulic behavior of the bioreactor. Recently, Sarkar and Chowdhury have studied the hydrodynamic behavior of PFBR with help of a dispersion model [16, 17] to study the non-ideal behavior of PFBR. As in their earlier work [6], they have used the growth study of *Lactobacillus sp.* for development of a mathematical model. The step and pulse studies revealed that real values of F(t) and E(t) deviated from the ideal values, and it was further enhanced with the increase of medium viscosity or concentration. A non-linear relationship was established between Pe and η_r for *Lactobacillus sp.* With increase of η_r, the value of Pe decreased which referred increase in non-ideal behavior. This study also revealed that with an increase of N_{Re} and volumetric flow rate, non-ideality of a PFBR increased. Hence, during the design of PFBR for a certain process, the effect of viscosity to be incorporated in model equation improved the performance of the reactor.

7.5 MATHEMATICAL MODELLING OF THE PACKED BED BIOREACTOR (PBBR)

PBBRs are used on a commercial scale, having several advantages like large surface area is available for growth of microorganisms and reduced residence time. The medium residence time is highly influential on the performance of PBBRs, as mass transfer between biofilm and working fluid depends on the flow rate. With progress of time, the thickness of the biofilm increases, which restricts the flow and mass transfer inside the reactor. Very little literature has been reported on RTD studies to determine hydrodynamic parameters for packed bed reactors [22–32]. Every research group has used different types of inert material and conducted RTD studies to measure different parameters of dispersion models [16, 17] for real bioreacting systems. However, the RTD experiment was carried out at a steady state condition, using packing material without considering growth of biofilm on packing material. Biofilm always reduces the path of the movement of the reaction medium; thus, it reduces the performance of the reactor. Therefore, such occurrence increases the non-ideality of PBBRs. Bernardez et al. [23] conducted the hydrodynamic study with help of a lab-scale experimental setup in the absence of any kind of biofilm. Such models are unable predict the actual non-ideality for real PBBRs. Moreover, no mathematical modelling on PBBRs by incorporating microbial reaction kinetics and non-ideal parameters has been reported so far.

In 2012, De and Chowdhury [5] conducted an RTD study with biofilm and developed a model for PBBR using both microbial reaction kinetics and non-ideal parameters. They had used the dispersion [16, 17] and Monod models to develop a mathematical model to describe the non-ideality of PBBRs. The model was validated for different thicknesses of biofilm, and they have expressed the bed porosity of the reactor as a function of thickness of biofilm. It has been observed that non-ideal plug flow occurred due to dispersion and stagnation in the bed. The present study also revealed that non-ideality increases with increase of biofilm thickness. Thus, during the design or modelling of a PBBR, thickness of biofilm should be considered as a performance-controlling parameter.

7.6 CONCLUSION

This study revealed that all kind of bioreactors behave as non-ideal systems due to the growth of biomass with time. As a result, the concentration of the medium and the thickness of biofilm also increase with time. It is essential to understand the hydrodynamic behavior of reactors through the RTD study. The parameters (viscosity, medium concentration, biofilm thickness, growth kinetics, rate constant) which are responsible for the performance of the reactor are to be included in mathematical modelling of the bioreactor. The deviation from ideal behavior is to be incorporated during the design of the bioreactor. Overall, this chapter is a valuable addition towards the study of non-ideality and design of different kinds of bioreactors.

REFERENCES

1. Al-Asheh S., Abu-Jdayil B., Abunasser N. and Barakat A., Rheological characteristics of microbial suspensions of *Pseudomonas aeruginosa* and *Bacillus cereus*. *Int. J. Biol. Macromol.*, **30**, 67–74 (2002).
2. Zhang Y.-H., Wang H.-Q., Liu S., Yu J.-T. and Zhong J.-J., Regulation of apparent viscosity and O2 transfer coefficient by osmotic pressure in cell suspensions of Panax notoginseng. *Biotechnol. Lett.*, **19**, 943–945 (1997).
3. Ajbar A., Study of the operability of nonideal continuous bioreactors. *Chem. Eng. Comm.*, **198**, 385–415 (2011).
4. Pedersen A. G., Bundgaard-Nielsen M., Nielsen J. and Villadsen J., Rheological characterization of media containing *Penicillium chrysogenum*. *Biotechnol. Bioeng.*, **41**, 162–164 (1993).

5. De R. and Chowdhury R., Hydrodynamics of a packed bed biofilm reactor (PBBR) for the removal of Hg^{2+} ion – RTD experiments with biotic and characteristically similar abiotic films and axial dispersion model. *J. Chem. Technol. Biotechnol.*, **88(9)**, 1612–1621 (2013).

6. Sarkar S., Chowdhury R. and Mukherjee A., Mathematical modelling of ideal and non-ideal continuous stirred tank bioreactor using simulated solution. *J. Chem. Technol. Biotechnol.*, **90(3)**, 484–491 (2015).

7. Sarkar S. and Chowdhury R., Simulation on non-ideal behaviour of plug flow bioreactor based on the growth study of *Lactobacillus* species. *Indian Chem. Eng.*, **64(4)**, 379–389, doi: 10.1080/00194506.2021.2008275 (2022).

8. Dutta S., Chowdhury R. and Bhattacharya P., Stability and response of bioreactor: An analysis with reference to microbial reduction of SO_2. *Chem. Eng. J.*, **133**, 343–354 (2007).

9. Al-Asheh S., Abu-Jdayil B., Abunasser N. and Barakat A., Rheological characteristics of microbial suspensions of *Pseudomonas aeruginosa* and *Bacillus cereus*. *Int. J. Biol. Macromol.*, **30**, 67–74 (2002).

10. Zhang Y.-H., Wang H.-Q., Liu S., Yu J.-T. and Zhong J.-J., Regulation of apparent viscosity and O_2 transfer coefficient by osmotic pressure in cell suspensions of Panax notoginseng. *Biotechnol. Lett.*, **19**, 943–945 (1997).

11. Martin A. D., Interpretation of residence time distribution data. *Chem. Eng. Sci.*, **55**, 5907–5917 (2000).

12. Melo P. A., Pinto J. C. and Biscaia E. C., Characterization of the residence time distribution in loop reactors. *Chem. Eng. Sci.*, **56**, 2703–2713 (2001).

13. Sanchez A., Valero F., Lafuente J. and Sola C., Highly enantioselective esterification of racemic ibuprofen in a packed bed reactor using immobilised *Rhizomucor miehei* lipase. *Enzyme Microb. Technol.*, **27**, 157–166 (2000).

14. Battaglia A., Fox P. and Pohland F. G., Calculation of residence time distribution from tracer recycle experiments. *Wat. Res.*, **27(2)**, 337–341 (1993).

15. Newell B., Bailey J., Islam A., Hopkins L. and Lant P., Characterising bioreactor mixing with residence time distribution (RTD) tests. *Water Sci. Technol.*, **37**, 43–47 (1998).

16. Fogler H. S., *Elements of chemical reaction engineering*, 2nd edn. Prentice-Hall of India Pvt. Ltd. (1992).

17. Levenspiel O., *Chemical reaction engineering*, 3rd edn. John Wiley & Sons, United States of America (1999).

18. Bailey J. E. and Ollis D. F., *Biochemical engineering fundamentals*, 32nd edn. McGraw Hill Inc. (1986).

19. Shuler M. L. and Kargi F., *Bioprocess engineering—Basic concepts*, Pearson Education, Inc. (2000).

20. Voloshin Y., Lawal A. and Panikov N. S., Continuous plug-flow bioreactor: Experimental testing with pseudomonas putida culture grown on benzoate. *Biotechnol. Bioeng.*, **91**, 254–259 (2005).

21. Limberg M. H., Pooth V., Wiechert W. and Oldiges M., Plug flow versus stirred tank reactor flow characteristics in two-compartment scale-down bioreactor: Setup-specific influence on the metabolic phenotype and bioprocess performance of *Corynebacterium glutamicum*. *Eng. Life Sci.*, **16**, 610–619 (2016).

22. Olivet D., Valls J., Gordillo M. A., Freixo A. and Sanchez A., Application of residence time distribution technique to the study of the hydrodynamic behaviour of a full-scale wastewater treatment plant plug-flow bioreactor. *J. Chem. Technol. Biotechnol.*, **80**, 425–432 (2005).

23. Bernardez L. A., Andrade Lima L. R. P. and Almeida P. F., The hydrodynamics of an upflow packed-bed bioreactor at low Reynolds number. *BJPG*, **2(3)**, 114–121 (2008).

24. Nardi I. R., Zaihat M. and Foresti E., Influence of the tracer characteristics on hydrodynamic models of packed-bed bioreactors. *Bioprocess Eng.*, **21**, 469–476 (1999).

25. Tembhurkar A. R. and Mhaisalkar V. A., Study of hydrodynamic behavior of a laboratory scale upflow anaerobic fixed film fixed bed reactor. *J. Environ. Sci. Eng.*, **48**, 75–80 (2006).
26. Borroto J. I., Dominguez J., Griffith J., Fick M. and Lecrec J. P., Technetium-99 m as a tracer for the liquid RTD measurement in opaque anaerobic digester: Application in a sugar wastewater treatment plant. *Chem. Eng. Process.*, **42**, 857–865 (2003).
27. Chua H., Hydrodynamics in the packed bed of anaerobic fixed filter reactor. *Water Sci. Technol.*, **33**, 1–6 (1996).
28. Wang Y., Sanly, Brannock M. and Leslie G., Diagnosis of membrane bioreactor performance through residence time distribution measurements-a preliminary study. *Desalination*, **236**, 120–126 (2009).
29. Camargo E., Canto C., Ratusznei S., Rodrigues J., Zaiat M. and Borzani W., Hydrodynamic analysis of sequencing batch biofilm reactor with liquid phase external circulation, *Interciencia*, **30**, 188–194 (2005).
30. Coelhoso I., Boaventura R. and Rodrigues A., Biofilm reactors: An experimental and modeling study of wastewater denitrification in fluidized-bed reactors of activated carbon particles. *Biotechnol. Bioeng.*, **40**, 625–633 (1992).
31. Sharvelle S., Mclamore E. and Banks M., Hydrodynamic characteristics in biotrickling filters as affected by packing material and hydraulic loading rate. *J. Environ. Eng.*, **134**, 346–352 (2008)
32. Seguret F., Racault Y. and Sardin M., Hydrodynamic behavior of full-scale trickling filters. *Water Res.*, **34**, 1551–1558 (2000).

8 Modeling and Simulating the Kinetics of Detoxification of Olive Mill Wastewater by Electrocoagulation Process

Reda Elkacmi and Rajaa Zahnoune

8.1 INTRODUCTION

Process modeling is based on the representation of a physical phenomenon using mathematical models to describe its behavior. It allows representing by mathematical models the various phenomena of transfer of mass, energy, and quantity of movement which occur in the various unit operations. The development of mathematical models and the use of computational languages have led to a better understanding of the kinetics of multi-component chemical reactions. On an industrial scale, process simulation is used to calculate the size of equipment, the quantity of energy required, the total yield, and the amount of waste produced.

Over the past years, the olive oil industry has been very important in Mediterranean countries, both in terms of wealth and tradition, with a total production of olives for olive oil in the EU of 12.9 million tons in 2018 [1]. This sector is considered as one of the largest consumers of water during different production steps, which causes a large production of liquid waste called olive mill wastewater (OMW). This liquid discharge generally has an intense red color, an acidic pH (3–6), high chemical oxygen demand (COD) (80–200 g/L), biological oxygen demand (BOD5) (12–60 g/L), and phenolic compounds (0.5–24 g/L) [2, 3].

For environmental and legal reasons, this hazardous effluent must therefore be treated before its final discharge. This subject is particularly critical in Morocco, as well as in the broader Mediterranean region, given the rapid expansion of the olive oil industry.

Various detoxification processes have been developed to reduce the polluted lead of OMW. Some examples of these technologies include physical processes such as dilution, evaporation, sedimentation, filtration, centrifugation, thermic treatment [4–6], biological treatment, aerobic and anaerobic digestion [7, 8], and physico-chemical processes such as lime treatment, coagulation-flocculation, and electro-Fenton oxidation [9–11]. However, some of these methods have significant disadvantages, including expensive equipment and monitoring systems, high reagent, energy requirements,

DOI: 10.1201/9781003435181-8

generation of toxic sludge or other by-products that require disposal, and involvement of complicated procedures which are economically unfeasible.

Electrocoagulation (EC) processes, widely used in the field of water treatment, have shown their flexibility in removing different forms of pollution and pollutants. This technology is based on the principle of soluble anodes. It consists of imposing a current between two electrodes (iron or aluminum) immersed in an electrolyte to generate, in situ, ions (Fe^{2+}, Fe^{3+}, Al^{3+}), capable of producing a coagulant in a solution and causing coagulation-flocculation of the pollutants present in the wastewater. The main reactions involved with aluminum electrodes as an example are:

At the anode:

- Metal oxidation: $Al \longrightarrow Al^{3+} + 3e^-$
- Hydrogen formation: $2Al + 3OH^- \longrightarrow Al_2O_3 + 3/2\, H_2 + 3\, e^-$

At the cathode:

- Water reduction: $H_2O + e^- \longrightarrow 1/2\, H_2 + OH^-$

A comprehensive study of detoxification of OMW by EC process with different electrodes is reported in the available literature [12–14], but EC treatment of OMW with aluminum electrodes along with the kinetic study and modeling approach is rarely reported, like the one of Ghahrchi et al. [15] who reported a total degradation of COD using stainless steel and aluminum as a cathode and anode electrodes, respectively. Moreover, the kinetic study determined that the COD removal obeyed the pseudo-second-order and intraparticle diffusion (IPD) kinetic model.

It is worth mentioning that EC is the result of adsorption of the pollutant on the solid formed after prior electrochemical dissolution of the aluminum cathode. The set of steps that leads to a separation by EC is extremely complex since there is also a stage of capture of these microparticles by the hydrogen bubbles, then agglomeration toward the free surface to create the flocs.

Advances in recent research show that there is a complementarity between two approaches to determining the kinetics of the adsorption reaction.

The first is based on a global vision of the phenomenon; it was necessary to determine the kinetics of elimination of three parameters COD, polyphenols (Ph), and coloration. The kinetic constant is determined from the experimental curves of pollutant removals; a stirred tank reactor (STR) is therefore used for this study. The influence of the various process parameters, namely, contact time, current density, and pH on the kinetic constant was determined.

The second approach consists of characterizing the adsorption phenomenon of the system. A study of the adsorption isotherms was carried out to determine the models adapted to the experience.

This chapter endeavors to represent the EC mechanism's effect of OMW detoxification and assess the influence of operating parameters on removal efficiencies in order to define the kinetic detoxification model that can be applied in STR to predict operating time for realizing an effective OMW treatment.

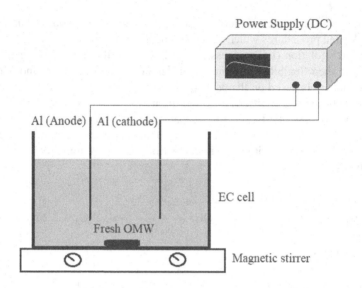

FIGURE 8.1   Schematic diagram of the experimental setup.

8.2   EXPERIMENTAL EC SETUP

EC experiments were performed in a laboratory-scale reactor made of plexiglass. Two aluminum electrodes with 12 cm × 5 cm × 0.2 cm dimensions, having a total effective surface area of 60 cm^2 and 1 cm electrode spacing, were arranged vertically. They were connected to terminals of a direct current power supply characterized by the ranges 0–6 A for current and 0–30V for voltage. The electrodes were immersed in 3.5 liters of fresh OMW under a constant agitation rate of 400 rpm (Figure 8.1).

8.3   KINETIC STUDY

As mentioned earlier, pollutant removal in the EC process occurred in two stages: the formation of metal flocs at the anode and the adsorption of cations on the surface of the flocs. Chemical kinetics are directly related to the rates of the chemical reactions involved, which depend on factors such as temperature and concentration.

To identify the models that best describe the kinetics of COD, PP, and dark color adsorption, first-order, second-order, pseudo-first-order, and pseudo-second-order models were evaluated. The fitting accuracy of kinetic data and models was tested by the regression coefficient R^2. The kinetic equilibriums were calculated and are summarized in Table 8.1. As can be observed, all R^2 values are too close to unity. Thus, first-order kinetic data can model the removal rate of color and pH while the COD follows the second-order model.

8.4   ADSORPTION EQUILIBRIUM ISOTHERMS

It should be pointed out that the experimental adsorption equilibrium isotherms are useful for describing the adsorption capacity of a specific adsorbent. Furthermore, the isotherm is critical for the analysis and design of adsorption systems, as well as

TABLE 8.1
Kinetic Constants and Regression Coefficients for the Removal of Color, pH, and COD at 25 and 66.66 mA/cm² of Current Densities

		First-Order Model		Second-Order Model		Pseudo-First-Order		Pseudo-Second-Order	
Parameters	CD (mA/cm²)	k (min⁻¹)	R²	k (min⁻¹)	R²	k (min⁻¹)	R²	k (min⁻¹)	R²
Color	25	0.0300	0.9492	0.0082	0.9336	0.0940	0.8640	0.0645	0.9360
	66.66	0.0728	0.9465	0.1033	0.8885	0.0670	0.9315	0.4763	0.8669
Ph	25	0.0260	0.8695	0.0023	0.6830	0.0222	0.6027	0.0243	0.8829
	66.66	0.0558	0.9728	0.0245	0.9357	7.4686	0.9170	0.1230	0.8773
COD	25	0.0184	0.9417	0.0011	0.9433	0.0265	0.8572	0.0153	0.9713
	66.66	0.0227	0.8736	0.0020	0.9360	0.0533	0.9732	0.0284	0.9837

Source: Reference [16].

the prediction of the appropriate models. Several theories of adsorption equilibrium were tested for the analysis and description of equilibrium adsorption data.

Two general-purpose models and a modified combined model were used to fit the experimental data: (1) the Langmuir model Eq. 8.1, (2) the Freundlich model Eq. 8.2, and (3) the Langmuir-Freundlich model Eq. 8.3 [16]:

$$q_e = \frac{q_{max}k_L C_e}{1 + k_L C_e} \tag{8.1}$$

$$q_e = k_F C_e^{\frac{1}{p}} \tag{8.2}$$

$$q_e = \frac{q_{max}k_{LF} C_e^n}{1 + k_{LF} C_e^n} \tag{8.3}$$

where q_e is the amount adsorbed at equilibrium (gram of adsorbate per gram of adsorbent); q_{max} represents the maximum adsorption capacity (gram of adsorbate per gram of adsorbent); C_e is the equilibrium concentration of the adsorbate (g/l); n is the index of heterogeneity; and k_L, k_F, and k_{LF} are the Langmuir, Freundlich, and Langmuir-Freundlich constants, respectively.

The three adsorption models were examined, keeping all other experimental conditions constant (pH = 5.2, V = 3.5 l for a treatment time of 45 min). Based on the linear forms of the models, the Langmuir and Freundlich parameters were calculated by graphing Ce/qe versus Ce and ln (qe) against ln Ce, respectively. qe was directly plotted versus Ce for the Langmuir-Freundlich model, and the three parameters (q_{max}, k_{LF}, and n) were found using the nonlinear regression approach. Table 8.2 summarizes all of the coefficients of the three models, allowing the conclusion that the Langmuir-Freundlich model best fits the experimental results of the reduction of color ($R^2 = 0.993$), pH ($R^2 = 0.992$), and COD ($R^2 = 0.991$).

TABLE 8.2
Langmuir, Freundlich, and Langmuir-Freundlich Isotherm Parameters for Color, pH, and COD Adsorption

Parameters	Adsorption parameters	Langmuir	Frendlish	Langmuir-Frendlish
Color	q_{max} (g/g)	0.534	–	–
	k_L (g/l)	12.07	–	–
	k_F (g/l)	–	1.304	–
	k_{LF} (g/l)	–	–	0.029
	R^2	0.927	0.903	0.993
pH	q_{max} (g/g)	3.507	–	–
	k_L (g/l)	0.717	–	–
	k_F (g/l)	–	1.469	–
	k_{LF} (g/l)	–	–	0.264
	R^2	0.481	0.844	0.992
COD	q_{max} (g/g)	9.708	–	–
	k_L (g/l)	0.075	–	–
	k_F (g/l)	–	0.019	–
	k_{LF} (g/l)	–	–	0.249
	R^2	0.790	0.943	0.991

Source: Reference [16].

8.5 VARIABLE ORDER KINETIC APPROACH

The kinetics of OMW treatment by EC tools should be examined to estimate the time required for detoxification. In order to specify the kinetics of fluoride removal, Hu et al. [17] first proposed a variable order kinetic (VOK) based on the Langmuir equation.

The VOK mathematical model was created to adequately represent and understand the adsorption kinetics, taking into consideration the evolution of phenomena over time related to changes in the mass of the adsorbent.

As shown in the previous section, floc formation and Al(OH)₃ precipitation followed first-order kinetics for decolorization and pH reduction and pseudo-second order for COD abatement.

It is evident from the published data that the treatment rate in the VOK approach is related to the aluminum release kinetics expressed by the total aluminum concentration in the solution $[Al]_{Tot}$.

$$-\frac{dC}{dt} = \varnothing_{Al} q_e \frac{d[Al]_{Tot}}{dt} \qquad (8.4)$$

where $\varnothing_{Al}$ is the floc formation efficiency.

$[Al]_{Tot}$ represents the concentration of total aluminum released in solution by the anode, which can be determined from Faraday's law:

$$-\frac{d[Al]_{Tot}}{dt} = \varnothing_C \frac{I}{ZFV} \qquad (8.5)$$

where $\emptyset_C$ represents the current efficiency (faradaic efficiency), I is the applied current (A), Z is the valence of the electrode metal (Z = 3 for aluminum), F is the Faraday constant (96,500 C) and V is the volume of the OMW solution (m^3).

The following equation is produced by combining Eqs. 8.4 and 8.5.

$$-\frac{dC}{dt} = \emptyset_{Al}\emptyset_C q_e \frac{I}{ZFV} \tag{8.6}$$

Eq. 8.7 can be written by combining Eq. 8.6 and the Langmuir-Freundlich model (Eq. 8.3).

$$-\frac{dC}{dt} = \emptyset_{Al}\emptyset_C \frac{q_{max}k_{LF}C_e^n}{1 + k_{LF}C_e^n} \frac{I}{ZFV} \tag{8.7}$$

The first order rate constant (k_1) can be expressed as:

$$k_1 = \emptyset_{Al}\emptyset_C q_{max} \frac{I}{ZFV} \frac{k_{LF}C_e^{n-1}}{\left(1 + k_{LF}C_e^{n-1}\right)} \tag{8.8}$$

Using Eq. 8.7, the pseudo-second-order rate constant (k_2) can be written as:

$$k_2 = \emptyset_{Al}\emptyset_C q_{max} \frac{I}{ZFV} \frac{k_{LF}C_e^n}{\left(1 + k_{LF}C_e^n\right)(C - C_e)^2} \tag{8.9}$$

The retention time required (t_r) to achieve an acceptable residual level of pollutants (Ce) in the STR can be determined by integrating (Eq. 8.7):

$$t_r = \frac{ZFV}{\emptyset_{Al}\emptyset_C q_{max}}\left[(C_0 - C) + \frac{1}{k_{LF}(1-n)}\left(C_0^{n-1} - C_e^{1-n}\right)\right] \tag{8.10}$$

The model (VOK) derived from the Langmuir-Freundlich equation will be applied to the experimental data of the reduction of the pollutant load present in the OMW. The effect of current density was studied under optimal conditions (pH = 5.2 and t = 45 min).

Figure 8.2 illustrates the the variations in the concentration of pollutants during EC and the model (VOK) as a function of time at I = 1.5, 2, 2.5, 3, 3.5, and 4 A corresponding to the current densities of 25, 33.33, 41.66, 50, 58.33, and 66.66 mA/cm², respectively.

It may be observed here that the VOK model applies strongly to experimental data. This confirms the validity of this approach in modeling OMW detoxification under different current densities (Figure 8.2).

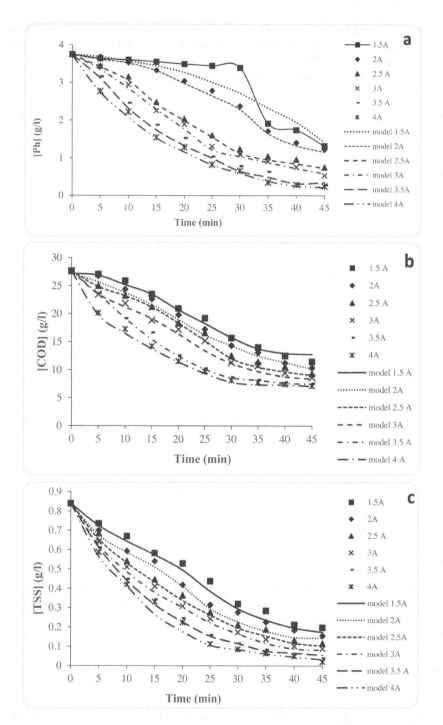

FIGURE 8.2 Variations of pollutant concentration during EC at different applied current densities and comparison with the predictions of the VOK model: (a) pH (b) COD (c) TSS [16].

8.6 CONCLUSION

In the present work, the EC mechanisms effect of OMW treatment is studied in order to develop a kinetic model to simulate the detoxification of OMW with EC process using bipolar aluminum electrodes in STR, based on the Langmuir-Freundlich adsorption model. The results showed good agreement between the predictive equation and the experimental data.

The development of a model describing the kinetics of detoxification of vegetable waters will allow the study to progress to the simulation of reactors on a pilot scale in order to predict their performance under various disturbances and conditions. This will allow, on the one hand, getting closer to industrial standards, and on the other hand to make any reactor of this study easily usable to carry out the depollution of various water impurities and to treat several types of wastewaters.

REFERENCES

1. Donner, M., & Radić, I. (2021). Innovative circular business models in the olive oil sector for sustainable Mediterranean agrifood systems. *Sustainability, 13*(5), 2588.
2. Rahmanian, N., Jafari, S. M., & Galanakis, C. M. (2014). Recovery and removal of phenolic compounds from olive mill wastewater. *Journal of the American Oil Chemists' Society, 91*(1), 1–18.
3. Elkacmi, R., & Bennajah, M. (2019). Advanced oxidation technologies for the treatment and detoxification of olive mill wastewater: A general review. *Journal of Water Reuse and Desalination, 9*(4), 463–505.
4. Masi, F., Bresciani, R., Munz, G., & Lubello, C. (2015). Evaporation–condensation of olive mill wastewater: Evaluation of condensate treatability through SBR and constructed wetlands. *Ecological Engineering, 80*, 156–161.
5. Paraskeva, P., & Diamadopoulos, E. (2006). Technologies for olive mill wastewater (OMW) treatment: A review. *Journal of Chemical Technology & Biotechnology: International Research in Process, Environmental & Clean Technology, 81*(9), 1475–1485.
6. Gebreyohannes, A. Y., Mazzei, R., & Giorno, L. (2016). Trends and current practices of olive mill wastewater treatment: Application of integrated membrane process and its future perspective. *Separation and Purification Technology, 162*, 45–60.
7. González-González, A., & Cuadros, F. (2015). Effect of aerobic pretreatment on anaerobic digestion of olive mill wastewater (OMWW): An ecoefficient treatment. *Food and Bioproducts Processing, 95*, 339–345.
8. Tsigkou, K., Terpou, A., Treu, L., Kougias, P. G., & Kornaros, M. (2022). Thermophilic anaerobic digestion of olive mill wastewater in an upflow packed bed reactor: Evaluation of 16S rRNA amplicon sequencing for microbial analysis. *Journal of Environmental Management, 301*, 113853.
9. Aktas, E. S., Imre, S., & Ersoy, L. (2001). Characterization and lime treatment of olive mill wastewater. *Water Research, 35*(9), 2336–2340.
10. Esteves, B. M., Rodrigues, C. S., Maldonado-Hódar, F. J., & Madeira, L. M. (2019). Treatment of high-strength olive mill wastewater by combined Fenton-like oxidation and coagulation/flocculation. *Journal of Environmental Chemical Engineering, 7*(4), 103252.
11. Ltaïef, A. H., Gadri, A., Ammar, S., Fernandes, A., Nunes, M. J., Ciríaco, L., Pacheco, M. J., & Lopes, A. (2019). Olive mill wastewater treatment by electro-Fenton with heterogeneous iron source. In *Wastes: Solutions, Treatments and Opportunities III* (pp. 333–338). CRC Press.

12. Ntaikou, I., Antonopoulou, G., Vayenas, D., & Lyberatos, G. (2020). Assessment of electrocoagulation as a pretreatment method of olive mill wastewater towards alternative processes for biofuels production. *Renewable Energy, 154*, 1252–1262.

13. Elkacmi, R., Boudouch, O., Hasib, A., Bouzaid, M., & Bennajah, M. (2020). Photovoltaic electrocoagulation treatment of olive mill wastewater using an external-loop airlift reactor. *Sustainable Chemistry and Pharmacy, 17*, 100274.

14. Salameh, W. K. B., Ahmad, H., & Al-Shannag, M. (2015). Treatment of olive mill wastewater by electrocoagulation processes and water resources management. *International Journal of Environmental and Ecological Engineering, 9*(4), 296–300.

15. Ghahrchi, M., Rezaee, A., & Adibzadeh, A. (2021). Study of kinetic models of olive oil mill wastewater treatment using electrocoagulation process. *Desalination and Water Treatment, 211*, 123–130.

16. Elkacmi, R., Kamil, N., & Bennajah, M. (2017). Upgrading of Moroccan olive mill wastewater using electrocoagulation: Kinetic study and process performance evaluation. *Journal of Urban and Environmental Engineering, 11*(1), 30–41.

17. Hu, C. Y., Lo, S. L., & Kuan, W. H. (2007). Simulation the kinetics of fluoride removal by electrocoagulation (EC) process using aluminum electrodes. *Journal of Hazardous Materials, 145*(1–2), 180–185.

9 Dynamics of Coexistence of Mixed Consortia in Sweet Meat Casein Whey
Modeling and Experimental

Arijit Nath, Sudip Chakraborty,
Madhumita Maitra, and Stefano Curcio

NOMENCLATURE

c_{lac}	Lactose concentration in abiotic phase	[M]
c_{lac_0}	Lactose concentration in abiotic phase at initial moment	[M]
c_{LA}	Lactic acid concentration in abiotic phase	[M]
c_{Eth}	Ethanol concentration in abiotic phase	[M]
c_{Rf}	Riboflavin concentration in abiotic phase	[M]
c_{s_i}	Concentration of any substrate in abiotic phase	[g L^{-1}]
$K_{LA_{lac}}$	Inhibition constant of lactic acid in presence of lactose	[M]
$K_{LA_{Rf}}$	Inhibition constant of lactic acid in presence of riboflavin	[M]
$K_{S_{LAB_{lac}}}$	Monod saturation constant of *Lactobacillus casei* (*JUCHE 2*) for lactose	[M]
$K_{S_{LAB_{Rf}}}$	Monod saturation constant of *Lactobacillus casei* (*JUCHE 2*) for riboflavin	[M]
$K_{S_{SAC_{lac}}}$	Monod saturation constant of *Saccharomyces cerevisae* (*JUCHE 3*) for lactose	[M]
$K_{S_{SAC_{Rf}}}$	Monod saturation constant of *Saccharomyces cerevisae* (*JUCHE 3*) for riboflavin	[M]
P_i	Concentration of any product in abiotic phase	[g L^{-1}]
t	Time	[h]
X_i	Concentration of any type consortium	[g L^{-1}]
X_{LAB}	Biomass concentration of *Lactobacillus casei* (*JUCHE 2*)	[g L^{-1}]
X_{LAB_0}	Biomass concentration of *Lactobacillus casei* (*JUCHE 2*) at initial moment	[g L^{-1}]
X_{SAC}	Biomass concentration of *Saccharomyces cerevisae* (*JUCHE 3*)	[g L^{-1}]
X_{SAC_0}	Biomass concentration of *Saccharomyces cerevisae* (*JUCHE 3*) at initial moment	[g L^{-1}]
$Y_{Eth/S_{lac}}$	Yield coefficient for ethanol production based on lactose	[Dimensionless]

DOI: 10.1201/9781003435181-9

$Y_{LA/X_{LAB}}$ Yield coefficient for lactic acid production based on   [Dimensionless]
            Lactobacillus casei (JUCHE 2)

$Y_{LA/S_{lac}}$ Yield coefficient for lactic acid production based on   [Dimensionless]
            lactose

$Y_{Rf/S_{lac}}$ Yield coefficient for riboflavin production based on   [Dimensionless]
            lactose

$Y_{X_{LAB}/S_{lac}}$ Yield coefficient for generation of *Lactobacillus casei*   [Dimensionless]
            (JUCHE 2) based on lactose

$Y_{X_{SAC}/S_{lac}}$ Yield coefficient for generation of *Saccharomyces*   [Dimensionless]
            cerevisae (JUCHE 3) based on lactose

$Y_{X_{LAB}/S_{Rf}}$ Yield coefficient for generation of *Lactobacillus casei*   [Dimensionless]
            (JUCHE 2) based on riboflavin

$Y_{X_{SAC}/S_{Rf}}$ Yield coefficient for generation of *Saccharomyces*   [Dimensionless]
            cerevisae (JUCHE 3) based on riboflavin

α_{Eth} Constant for Luedeking Piret model for ethanol production [Dimensionless]
α_{Rf} Constant for Luedeking Piret model for riboflavin   [Dimensionless]
            production

β_{Eth} Constant for Luedeking Piret model for ethanol production   [h⁻¹]
β_{Rf} Constant for Luedeking Piret model for riboflavin production [h⁻¹]
μ_{LAB} Specific growth rate of *Lactobacillus casei (JUCHE 2)*   [h⁻¹]
$\mu_{maxLABlac}$ Maximum specific growth rate of *Lactobacillus casei*   [h⁻¹]
            (JUCHE 2) based on lactose

$\mu_{maxLABRf}$ Maximum specific growth rate of *Lactobacillus casei*   [h⁻¹]
            (JUCHE 2) based on riboflavin

μ_{SAC} Specific growth rate of *Saccharomyces cerevisae*   [h⁻¹]
            (JUCHE 3)

$\mu_{maxSAClac}$ Maximum specific growth rate of *Saccharomyces cerevisae*   [h⁻¹]
            (JUCHE 3) based on lactose

$\mu_{maxSACRf}$ Maximum specific growth rate of *Saccharomyces cerevisae*   [h⁻¹]
            (JUCHE 3) based on riboflavin

9.1   INTRODUCTION

The importance of dynamics of mixed populations of microorganisms can be traced due to the response of many ecological systems to stress [1]. In all natural environments, microbes exist in potentially mixed populations [2]. It is very critical to understand the dynamics of mixed populations of microorganisms, their interactions with each other, and their coexistence in natural systems.

Mixed populations of microorganism were already used in waste treatment, specially, activated sludge process, hydrocarbon degradation, wastewater treatment, and various biochemical approaches like production of fermented food and nutraceuticals [3–24]. In natural systems, microbes may exhibit commensalism, mutualism, antagonism, and synergism. Due to the availability and nature of substrates, consortia can be exhibited either simultaneously or individually. In cases of antagonism, one group of consortia is inhibited by others whereas mutualism offers positive influence on simultaneous growth of consortia.

Dairy industries are a major economic source of tropical and subtropical countries, and are generating large volumes of waste liquid effluent, namely, whey. Whey is a heterogeneous mixture of different biomolecules such as lactose, proteins, i.e., bovin serum albumin, α-lactalbumin, β-globulin etc., which offer the adequate nutrients to different flora, and fauna [25]. During the abundant growth of different lactic acid, bacteria and yeast in sweet meat casein whey at aerobic or anaerobic situations, different value-added metabolites are synthesized, which play as promoting, inhibitory or may be ineffective on the growth of consortia. Therefore, it is obviously a unique attempt to understand the dynamics of coexistence of mixed consortia in sweet meat casein whey. The results obtained in this study would be beneficial to the scientific community and industries dealing with disposal of casein whey.

The objective of this chapter is to develop a deterministic, unstructured mathematical model regarding microbe-microbe interaction in sweet meat casein whey. Two facultative aerobic strains, namely *Lactobacillus casei (JUCHE 2)* and *Saccharomyces cerevisae (JUCHE 3)*, were isolated by present research from sweet meat casein whey. Simulated casein whey containing lactose concentration 50 g L^{-1} has been used to develop the model equations. The deterministic model has been developed for a small batch-type bioreactor (Erlenmeyer flask) using differential mass balance equations of different components and the model parameters. From the preliminary laboratory batch experiment, it was observed that the consortia can assimilate both lactose and riboflavin for their growth. During the growth, ethanol, and extracellular secondary metabolite riboflavin are synthesized by *Saccharomyces cerevisae (JUCHE 3)*. The growth of *Saccharomyces cerevisae (JUCHE 3)* is inhibited by synthesized ethanol, whereas riboflavin is utilized by both consortia for their growth. Lactic acid is synthesized by *Lactobacillus casei (JUCHE 2)* as a byproduct, which also inhibits the its own growth. Microbial growth was inhibited at high lactose concentration, i.e., 50 g L^{-1}. The model parameters, namely, maximum specific growth rate, saturation constant, and yield coefficient have been determined through analytical methods and regression analysis of data obtained from programmed batch experiments in Erlenmeyer flasks. Combinations of statistical and deterministic methodologies have been adopted. In a later exercise, the model has been validated by comparison of predicted data with experimental results obtained from a large 5L (working volume 2 L) bioreactor using casein whey as a growth medium.

9.2 MATERIALS AND METHODS

9.2.1 MATERIALS

Highly purified whey proteins, such as, α-lactalbumin, bovin serum albumin, IgG, IgA, IgM, lactoperoxidase, and lactoferrin were procured from Sigma Aldrich; USA. Lactose, riboflavin, tryptone, yeast extract, and beef extract, lactic acid were procured from HIMEDIA, Mumbai, India. di-potassium hydrogen phosphate, sodium acetate, manganese sulfate, magnesium sulfate, and methanol were procured from Merck, Mumbai, India. Ammonium citrate, hydrochloric acid, and sodium hydroxide were procured from Ranbaxy, Mumbai, India. Sweet meat casein whey was collected from Hindusthan sweet, Kolkata, India.

9.2.2 EQUIPMENT

A BOD incubator cum shaker (Sartorius AG, Göttingen, Germany), an indigenous microfiltration unit along with a cellulose acetate membrane of 47 mm diameter and 0.45 μm pore size, a water bath (Sarada Chemicals, Kolkata, India), a UV laminar flow hood, a hot air oven (Bhattacharya & Co., Kolkata, India), a magnetic stirrer, a cold centrifuge (C-24) (Remi Instruments Ltd., Mumbai, India), an autoclave (G.B. Enterprise, Kolkata, India), and an Arium 611DI ultrapure water system (Sartorius AG, Göttingen, Germany) were used. A 5-L glass-made bioreactor (working volume 2 L) associated with capillary niddle and a suction tube were used.

9.2.3 ANALYTICAL INSTRUMENTS

A digital pH meter, a digital weighing machine (Sartorius AG, Göttingen, Germany), a high-performance liquid chromatography (HPLC) (Perkin Elmer, Series 200) along with an RI detector, a UV-visible diode-array detector, a double-beam UV–vis spectrophotometer (SPECTRASCAN UV 3600-Chemito), and a field emission scanning electron microscope (FESEM) (Model: JSM 6700F, Jeol Ltd., Japan) were used. For carbohydrate, lactic acid, and ethanol analysis, a BIO-RAD Aminex® Fermentation Monitor column (150 × 7.8 mm, 5 μm), and for riboflavin estimation reversed-phase Discovery C_{18} column (150 mm × 4.6 mm, 5 μm) were used. Sulfuric acid (0.001 M) was used as the mobile phase (0.8 mL min^{-1}) in isocretic mode, and the temperature of the HPLC was maintained at 330 K for estimation of carbohydrates, lactic acid, and ethanol. A potassium di-hydrogen phosphate buffer (0.1 mol L^{-1}, pH 7.0), and methanol (v:v: : 9:1) was used as the mobile phase (0.7 mL min^{-1}) in isocratic mode, and temperature of the HPLC was maintained at 298 K for estimation of riboflavin. The colorimetric measurement by spectrophotometer was used to determine concentrations of biomass through the measurement of optical density of the samples withdrawn from the large and small bioreactors.

9.2.4 MICROORGANISM

Two isolated facultative aerobic strains, namely, *Lactobacillus casei (JUCHE 2)* and *Saccharomyces cerevisae (JUCHE 3)*, were used in the present investigation. The isolated consortia had been isolated from sweet meat whey, and were identified by MTCC (Microbial Type Culture Collection & Gene Bank, of Institute of Microbial Technology, Sector 39-A, Chandigarh-160036) through 16s rRNA analysis. Both the consortia were facultative in nature; therefore, the anaerobic condition (without agitation) was maintained during all experiments described in the following section.

9.2.5 GROWTH MEDIUM PREPARATION

Simulated casein whey and real casein whey were used as microbial growth medium. The simulated casein whey contained per liter: β-lactoglobulin 3.5 g, α-lactalbumin

1.4 g, bovin serum albumin 0.4 g, IgG 0.2 g, IgA 0.3 g, IgM 0.1 g, lactoperoxidase 0.06 g, lactoferrin 0.05 g, ammonium citrate 2.0 g, sodium acetate 5.0 g, magnesium sulfate 0.1 g, manganese sulfate 0.5 g, dipotassium phosphate 2.0 g, and lactose 40.0 g. pH of the medium was adjusted at 7.0 by 0.1 N sodium hydroxide and 0.1 N hydrochloric acid. Sterilization of all components of the growth medium except carbohydrates was done in an autoclave at 121°C for 15 minutes. Lactose and casein whey were sterilized using an indigenous microfiltration unit equipped with a cellulose acetate membrane (47 mm diameter, pore size 0.45 μm) for its sensitivity toward high temperature.

9.2.6 INOCULUM PREPARATION

Pre-culture of the strain *Lactobacillus casei (JUCHE 2)* and *Saccharomyces cerevisae (JUCHE 3)* were performed using simulated casein whey as a growth medium. During the experiment, 37°C incubation temperature for *Lactobacillus casei (JUCHE 2)* and 32°C incubation temperature for *Saccharomyces cerevisae (JUCHE 3)* were maintained. During the pre-culture period, one day for *Lactobacillus casei (JUCHE 2)*, and 3 days for *Saccharomyces cerevisae (JUCHE 3)* were fixed based on sufficient growth indicated by OD_{660}. Both the consortia were preserved by sterile glycerol with liquid growth medium (v:v: : 2:1). Adaptations of the cultures *Lactobacillus casei (JUCHE 2)* and *Saccharomyces cerevisae (JUCHE 3)* to simulated sweet meat casein whey were performed by repetitive subculturing [26]. Prior to the experiment repeated three times subculture were conducted for two consortia separately in Erlenmeyer flasks. Last adopted cultures were stored for use in the experiments conducted in the Erlenmeyer flasks as well as in the lab scale bioreactor.

9.2.7 GROWTH STUDIES IN SMALL REACTORS

Experiments were carried out in small reactors (Erlenmeyer flasks) with a working volume of 50 mL each at anaerobic condition (without agitation). The deterministic unstructured mathematical model and model parameters were evaluated using simulated sweet meat casein whey as a growth medium. Different series of experiments were conducted to determine the growth kinetics based on either lactose or riboflavin as the limiting substrates in the simulated whey medium. Individual experiments were conducted with the aid of evaluating different kinetic parameters using monoculture. Simulated whey medium using initial lactose concentration ranging from 0.4–1.2% (w/v) were used to determine the growth kinetic parameters and yield coefficients based on lactose as the limiting substrate. Instead of lactose, riboflavin was used in the same concentration range, i.e., 0.4–1.2% (w/v) to evaluate the growth kinetic parameters and yield coefficients based on riboflavin. Experiments were also conducted by addition of different concentrations of either lactic acid or ethanol along with either lactose or riboflavin in the simulated whey medium. Inoculation was done with adapted culture, and the inoculum size was maintained at 4.0% (w/v).

Experiments were also conducted by mixed consortia using simulated whey medium by batch time progress. In this case, equal inoculum sizes of two consortia were considered, and total inoculum size was maintained at 4.0% (w/v). The average environmental temperatures of tropical and subtropical countries are 30–40°C. Generally, pH of sweet meat casein whey is varied, ranging from 4.5–6. Therefore, experiments were conducted at a fixed incubation temperature of 32°C and pH 6. For each experiment, samples were withdrawn at 2-hour intervals, and biomass growth, substrate utilization, and product formation were estimated. The process was conducted up to 24 hours after which cells enter the stationary phase of growth.

9.2.8 EXPERIMENTS IN A BENCH-TOP REACTOR

Batch experiments were also carried out in a 5L bench-top reactor with a working volume of 2L. There was no lid associated with bioreactor. Prior to the experiment, sweet meat casein whey was used as a microbial growth medium. Inoculation was done (4.0% [w/v]) with adapted culture, and equal inoculum sizes of two consortia were maintained. The initial pH 6 of sweet meat casein whey was adjusted by 0.1 N hydrochloric acid and 0.1 N sodium hydroxide. An incubation temperature of 32°C was maintained by hot/cold water circulation and temperature controller. The experiment was conducted in laminar air flow hood for avoiding contamination by any other microbes. Samples were withdrawn from the bioreactor through a capillary needle into the evacuated sample tubes. The sample tubes were immediately placed in a refrigerator at 4°C for inactivation of cellular activities. The samples were analyzed for the determination of concentration of biomass, abiotic lactose, lactic acid, riboflavin, and ethanol following the methods described in Section 9.2.9.

9.2.9 ANALYTICAL METHODS

9.2.9.1 Estimation of Yield Coefficient

Yield coefficients of the cell, growing on either lactose or riboflavin, were determined using either the biomass or mass of lactic acid or mass of riboflavin or mass of ethanol were formed over a period of 2 hours, and the quantum of substrate were consumed over the same time period during the initial phase of exponential growth. Therefore,

$$Y_{c_{X_i}/c_{s_i}} = \frac{\Delta c_{X_i}}{|\Delta c_{s_i}|} \tag{9.1}$$

$$Y_{c_{P_i}/c_{s_i}} = \frac{\Delta c_{P_i}}{|\Delta c_{s_i}|} \tag{9.2}$$

Similarly, yield coefficients of byproduct lactic acid formation per unit biomass were determined using the mass of lactic acid were produced per quantum of biomass were produced over the same time period during the initial phase of exponential growth. Time difference was considered to be 2 hours. Therefore,

$$Y_{c_{P_i}/c_{X_i}} = \frac{\Delta c_{P_i}}{\Delta c_{X_i}} \tag{9.3}$$

9.2.9.2   Estimation of Constants of the Luedeking Piret Model

Constants of the Luedeking Piret model have been determined by plotting $\left(\dfrac{d[c_{LA}]}{[c_x]dt} \right)$

versus μ. The quantum of product formation with respect to biomass formed over a period of 2 hours during the initial phase of exponential growth was used to determine the constant of Eq. 9.3.

9.2.9.3   Cell Concentration Determination

Individual cell growths of proposed consortia were determined by plate count method. Briefly, Man–Rogosa–Sharpe (MRS) agar plate and Sabouraud Dextrose Broth (SDB) agar plate were used for growth of *Lactobacillus casei (JUCHE 2)* and *Saccharomyces cerevisae (JUCHE 3)* respectively. Cultures were diluted appropriately, and spread plate technique was employed. A constant incubation temperature of 37°C, and 24 hours of incubation were maintained for cell growth. A linear correlation between colony-forming unit (CFU) and biomass concentration was adopted. Biomass concentrations were also estimated by colorimetric and dry cell weight method.

9.2.9.4   Estimation of Carbohydrate, Lactic Acid, Ethanol, and Riboflavin Concentration

The supernatant obtained after centrifugation of 20 mL culture broth was analyzed using the high-pressure liquid chromatography (HPLC) to determine the concentrations of lactose, lactic acid, ethanol, and riboflavin in the abiotic phase. Centrifugation was done at 10,000 rpm and 4°C for 15 minutes.

9.2.9.5   FESEM Analysis

Harvested broth was collected at different time intervals and placed on glass cover slips individually. Samples were dried at vacuum condition and coated by platinum for 120 seconds at 10 mA current and 10 kV. Coated samples were placed in scanning electron microscopy (SEM) for analysis. In field emission scanning electron microscopy (FESEM), working distance 8 mm, 5 kV, 10 mA current, and secondary electron flow were used.

9.3   THEORETICAL ANALYSIS

The deterministic unstructured non-segregated mathematical model has been developed by some assumptions.

9.3.1   Assumptions

1. Two consortia, *Lactobacillus casei (JUCHE 2)* and *Saccharomyces cerevisae (JUCHE 3)*, grow in sweet meat casein whey considering lactose as a nutrient source under uncontrolled pH.
2. Extracellular secondary metabolite riboflavin is synthesized by *Saccharomyces cerevisae (JUCHE 3)*, which is utilized by both consortia.

3. Lactic acid is synthesized by *Lactobacillus casei (JUCHE 2)* as a byproduct, which inhibits its own growth. Growth of *Saccharomyces cerevisae
(JUCHE 3)* and synthesis of ribovlavin and ethanol are not affected by pH
of growth medium.

4. Ethanol is synthesized by *Saccharomyces cerevisae (JUCHE 3)*, which also
inhibits its own growth.

5. Photograph of growth medium by field emission scanning microscope
at different time interval it was observed that only *Lactobacillus casei
(JUCHE 2)* and *Saccharomyces cerevisae (JUCHE 3)* were present in
growth medium. Therefore, it is assumed that the growth medium is
not contaminated by other microbial consortia during the experimental
period.

9.3.2 GROWTH KINETICS OF *LACTOBACILLUS CASEI (JUCHE 2)*

Growth rate of *Lactobacillus casei (JUCHE 2)* depends upon two substrates, namely
lactose and secondary metabolite riboflavin. At high substrate concentration in the
microbial growth medium, i.e., 50 g L^{-1}, its growth is inhibited by both lactose and
lactic acid. Therefore,

$$\mu_{LAB} = \left(\frac{\mu_{\max LAB_{lac}} \cdot c_{lac}}{K_{sLAB_{lac}} + c_{lac}} + \frac{\mu_{\max LAB_{Rf}} \cdot c_{Rf}}{K_{sLAB_{Rf}} + c_{Rf}} \right) \left(1 - \frac{c_{lac}}{c_{LAB_{lac}}^{*}} \right)^{m} \left(1 - \frac{c_{LA}}{c_{LA}^{*}} \right)^{n} \quad (9.4)$$

From the experimental results it was observed that at lactose concentration 100 g
L^{-1} in the microbial growth medium, there was no microbial growth. Therefore,
critical concentration of lactose ($c_{LAB_{lac}}^{*}$) in the microbial growth medium was considered 100 g L^{-1} for *Lactobacillus casei (JUCHE 2)*. Similarly, the concentration
of lactic acid, 100 g L^{-1} in microbial growth medium, growth of *Lactobacillus
casei (JUCHE 2)* was totally inhibited. Critical concentration of lactic acid
($c_{LAB_{lac}}^{*}$) in microbial growth medium was considered 100 g L^{-1} for *Lactobacillus
casei (JUCHE 2)*.

9.3.3 GROWTH KINETICS OF *SACCHAROMYCES CEREVISAE (JUCHE 3)*

Growth rate of *Saccharomyces cerevisae (JUCHE 3)* also depends upon both lactose and secondary metabolite riboflavin. High substrate concentration, i.e., 50 g L^{-1}
growth of *Saccharomyces cerevisae (JUCHE 3)* is inhibited. Synthesized ethanol
could also inhibit its own growth. Therefore,

$$\mu_{SAC} = \left(\frac{\mu_{\max SAC_{lac}} c_{lac}}{K_{sSAC_{lac}} + c_{lac}} + \frac{\mu_{\max SAC_{Rf}} c_{Rf}}{K_{sSAC_{Rf}} + c_{Rf}} \right) \left(1 - \frac{c_{lac}}{c_{lac}^{*}} \right)^{m} \left(1 - \frac{c_{Eth}}{c_{Eth}^{*}} \right)^{n} \quad (9.5)$$

From the experimental results, it was observed that growth of *Saccharomyces
cerevisae (JUCHE 3)* is totally inhibited at lactose concentration 100 g L^{-1}. Therefore,
critical concentration of lactose ($c_{SAC_{lac}}^{*}$) in the microbial growth medium was

considered 100 g L^{-1} for *Saccharomyces cerevisae (JUCHE 3)*. The concentration of ethanol, 100 g L^{-1} in the microbial growth medium, growth of *Saccharomyces cerevisae (JUCHE 3)* was totally inhibited. Therefore, critical concentration of lactic acid (c_{LA}^{*}) in the microbial growth medium was considered 100 g L^{-1} for *Saccharomyces cerevisae (JUCHE 3)*.

9.3.4 LACTIC ACID FORMATION KINETICS

Formation of lactic acid by *Lactobacillus casei (JUCHE 2)* is growth-associated. Therefore,

$$c_{LA} = Y_{LA/X_{LAB}} c_{LAB} \tag{9.6}$$

9.3.5 RIBOFLAVIN FORMATION KINETICS

Formation of riboflavin by *Saccharomyces cerevisae (JUCHE 3)* may described by the Luedeking Piret model, Therefore,

$$C_{Rf} = \alpha_{Rf} c_{SAC} + \beta_{Rf} \tag{9.7}$$

9.3.6 ETHANOL FORMATION KINETICS

Formation of ethanol by *Saccharomyces cerevisae (JUCHE 3)* may described by the Luedeking Piret model, Therefore,

$$c_{Eth} = \alpha_{Eth} c_{SAC} + \beta_{Eth} \tag{9.8}$$

9.3.7 LACTOSE UTILIZATION

Lactose utilization by two consortia is described by following equation:

$$c_{lac} = -\left(\frac{1}{Y_{c_{LAB}/c_{lac}}} + \frac{1}{Y_{c_{LA}/c_{lac}}} \right) c_{LAB} - \left(\frac{1}{Y_{c_{SAC}/c_{lac}}} + \frac{1}{Y_{c_{Eth}/c_{lac}}} + \frac{1}{Y_{c_{Rf}/c_{lac}}} \right) c_{SAC} \tag{9.9}$$

9.3.8 MASS BALANCE EQUATIONS

Different mass balance equations are described below.

$$\frac{dc_{LAB}}{dt} = \mu_{LAB} c_{LAB} \tag{9.10}$$

$$\frac{dc_{SAC}}{dt} = \mu_{SAC} c_{SAC} \tag{9.11}$$

$$\frac{dc_{LA}}{dt} = Y_{LA/X_{LAB}} \mu_{LAB} c_{LAB} \tag{9.12}$$

$$\frac{dc_{Rf}}{dt} = \left(\alpha_{Rf} \mu_{SAC} + \beta_{Rf} \right) c_{SAC} - \left(\frac{\mu_{LAB} c_{LAB}}{Y_{c_{LAB}/c_{Rf}}} + \frac{\mu_{SAC} c_{SAC}}{Y_{c_{SAC}/c_{Rf}}} \right) \tag{9.13}$$

$$\frac{dc_{Eth}}{dt} = \left(\alpha_{Eth} \mu_{SAC} + \beta_{Eth} \right) c_{SAC} \tag{9.14}$$

$$\frac{dc_{lac}}{dt} = -\left(\frac{1}{Y_{c_{LAB}/c_{lac}}} + \frac{1}{Y_{c_{LA}/c_{lac}}} \right) \mu_{LAB} c_{LAB} - \left(\frac{1}{Y_{c_{SAC}/c_{lac}}} + \frac{1}{Y_{c_{Eth}/c_{lac}}} + \frac{1}{Y_{c_{Rf}/c_{lac}}} \right) \mu_{SAC} c_{SAC} \tag{9.15}$$

The initial boundary conditions are as follows,
At t = 0

$$\begin{bmatrix} c_{LAB} = c_{LAB_0} \\ c_{SAC} = c_{SAC_0} \\ c_{LA} = 0 \\ c_{Rf} = 0 \\ c_{Eth} = 0 \\ c_{lac} = c_{lac_0} \end{bmatrix} \tag{9.16}$$

9.3.9 STATISTICAL ANALYSIS

The above mathematical model describes the dynamic behavior of different components of this system. The model involves sets of ODEs of the following form:

$$\frac{dC}{dt} = f(C, P) \tag{9.17}$$

where C is a vector of concentrations of eight intracellular and extracellular components under the consideration and P is a vector of model parameters, namely, $K_{LA}, K_{S_{LAB_{lac}}}, K_{S_{LAB_{Rf}}}, K_{S_{SAC_{lac}}}, K_{S_{SAC_{Rf}}}, Y_{c_{Eth}/c_{lac}}, Y_{c_{LA}/c_{LAB}}, Y_{c_{LA}/c_{lac}}, Y_{c_{Rf}/c_{lac}}, Y_{c_{LAB}/c_{lac}}, Y_{c_{LAB}/c_{Rf}}, Y_{c_{SAC}/c_{lac}}, Y_{c_{SAC}/c_{Rf}}, \alpha_{Eth}, \alpha_{Rf}, \beta_{Eth}, \beta_{Rf}, \mu_{\max LAB_{lac}}, \mu_{\max LAB_{Rf}}, \mu_{\max SAC_{lac}}$ and, $\mu_{\max SAC_{Rf}}$.

The model parameters, P, may be determined by two methods, which are as follows,

Method 1: Determination of the parameters through minimization of an objective function, including the sum of squared errors/deviations between the predicted and experimental values of concentrations of components under consideration.

Method 2: Determination of parameters through minimization of the objective function, including the sum of squared errors/differences between the predicted and experimental values of individual variables as obtained from the batch of experimental runs, dedicated to the parameter estimation.

Since no literature data is available on the possible ranges of values of the above-mentioned kinetic parameters under the mentioned experimental system, in the present case, Method 2 has been adopted to determine the parameters. Regression analyzes of data obtained from the parameter estimation experiments, described in Section 9.2.6 have been done. For each regression analysis, the correlation coefficient and confidence interval have been determined

9.3.9.1   Correlation Coefficient

In order to estimate the strength of respective analysis, the coefficient of correlation has been determined. For any regression equation $y = a + bz$, the coefficient of correlation, r^* may be represented as follows:

$$r^* = b \sqrt{\frac{n\sum\limits_{i=1}^{n} z_i^2 - \left(\sum\limits_{i=1}^{n} z_i\right)^2}{n\sum\limits_{i=1}^{n} y_i^2 - \left(\sum\limits_{i=1}^{n} y_i\right)^2}} \qquad (9.18)$$

where, i signifies any value, n represents sample size.

9.3.9.2   Confidence Interval

The quality of parameter estimation is indicated by the confidence interval. A smaller interval means a high quality of parameter estimate or a smaller error. A $100(1 - \alpha)$ percent prediction interval on new observations is obtained using the following expression,

$$y_{obs} = y_p \pm t_{\alpha/2,n-2} \sqrt{MS_E \left[1 + \frac{1}{n} + \frac{\left(z_p - \bar{z}\right)}{\sum\limits_{i=1}^{n}\left(z_i - \bar{z}\right)^2} \right]} \qquad (9.19)$$

$$MS_E = SSE/dof\,(SSE) \qquad (9.20)$$

where MS_E is the error mean square, dof is the degree of freedom, and SSE is the sum of square of deviations between observed and predicted values of the regressed variable used for parameter estimation.

To calculate the confidence interval for any value, z_p of independent variable, the deviation between the observed value, y_{obs} and predicted regressed value, y_p of dependent variable is calculated. Using the value of the deviation and the value

of $\sqrt{MS_E \left[1 + \dfrac{1}{n} + \dfrac{(z_p - \bar{z})}{\sum\limits_{i=1}^{n}(z_p - \bar{z})^2} \right]}$, the value of $\alpha/2$ is determined by consulting the

student's t distribution table. The percent confidence interval or prediction interval is given by $100(1 - \alpha)$. At x% confidence interval, $\alpha = (100 - x)/100$. The values of confidence intervals for regression equations involving model parameters have been determined using the new observations, which are not utilized for parameter estimation.

9.3.9.3   Estimation of Model Fits for the Lab-Scale Reactor
The accuracy and confidence level of the proposed model equations have been evaluated by MS_E criterion and Fisher's F ratio test.

$$F\ ratio = \frac{MS_R}{MS_E} \tag{9.21}$$

where, $MS_R = SSR/dof$, and $dof = 1$

$$SSR = \sum \left(y_{pi} - \bar{y} \right)^2 \tag{9.22}$$

and

$$MS_E = \frac{\left(y_{obs_i} - y_{p_i} \right)^2}{dof} \tag{9.23}$$

$$dof = n - m \tag{9.24}$$

where m is the number of constants involved in model equations.

The significance level, p, is determined from the value of F, dof of quantity in the numerator, i.e., 1, or the denominator, i.e., $(n-m)$. Evaluation of such confidence level, and accuracies of model equations (usually referred to as the strength of the mathematical model) using the statistical approach has been done by other pioneer investigators in this field.

Proposed model equations (Eqs. 9.10–9.15) substantiated by initial conditions (Eq. 9.16) have been solved using the fourth-order Runge-Kutta method with the aid of MATLAB 7.0. Validity of the model has been checked by comparing it with experimental data obtained from the lab-scale bench-top bioreactor.

9.4   RESULTS AND DISSCUSSION

This is a unique attempt, to the best of our knowledge, to develop the kinetic model of mixed consortia namely, *Lactobacillus casei (JUCHE 2)* and *Saccharomyces cerevisae (JUCHE 3),* growing in sweet meat casein whey at uncontrolled pH. In Figure 9.1, the coexistence of two proposed consortia in sweet meat casein whey under uncontrolled pH is depicted. The mathematical model was developed with respect to predict the transient variation of concentrations of different components in the system. The quality of the model fits, and the estimated model parameters are discussed in detail.

The deterministic model equations developed in the present investigation contain large numbers of parameters, whose values have mostly been determined experimentally. It may be mentioned that no adjustable parameters have been used for simulation purposes. The magnitudes of estimated parameters are shown in Table 9.1. The corresponding confidence intervals and the correlation indices are provided in Table 9.2. For all evaluated model parameters, the confidence intervals are checked for new observations of respective experiments dedicated to the evaluation of parameters. New observations are independent of those used to obtain the values of model parameters through regression analysis. In most cases confidence intervals of 98% or more, and the correlation indices of 0.98 to 0.99 have been obtained.

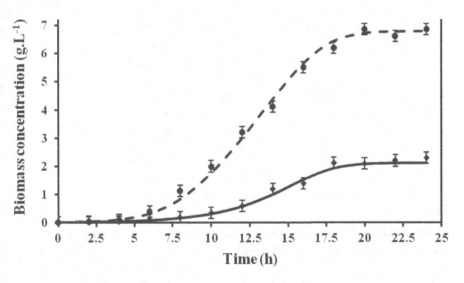

FIGURE 9.1   Simulated (lines) and experimental (points) time histories of biomass concentration of *Lactobacillus casei (JUCHE 2)* and *Saccharomyces cerevisae (JUCHE 3)* in the growth medium. Lines: biomass concentration of *Lactobacillus casei (JUCHE 2)* (–), biomass concentration of *Saccharomyces cerevisae (JUCHE 3)* (–); Points: Concentration of biomass of *Lactobacillus casei (JUCHE 2)* at simulated whey medium (■), concentration of biomass of *Lactobacillus casei (JUCHE 2)* at real casein whey medium (▲), concentration of biomass of *Saccharomyces cerevisae (JUCHE 3)* at simulated whey medium (•), and concentration of biomass of *Saccharomyces cerevisae (JUCHE 3)* at real casein whey medium *(JUCHE 3)* (♦).

TABLE 9.1

List of Parameters Used in Models

Parameter	Value	Reference
$K_{LA_{lac}}$	1.0×10^{-2} M	Evaluated
$K_{LA_{Rf}}$	6.0×10^{-3} M	Evaluated
$K_{S_{LAB_{lac}}}$	2.8×10^{-2} M	Evaluated
$K_{S_{LAB_{Rf}}}$	2.0×10^{-2} M	Evaluated
$K_{S_{SAC_{lac}}}$	4.2×10^{-2} M	Evaluated
$K_{S_{SAC_{Rf}}}$	3.2×10^{-2} M	Evaluated
$Y_{Eth/S_{lac}}$	0.5	Evaluated
$Y_{LA/X_{LAB}}$	0.45	Evaluated
$Y_{LA/S_{lac}}$	0.48	Evaluated
$Y_{Rf/S_{lac}}$	0.25	Evaluated
$Y_{X_{LAB}/S_{lac}}$	0.48	Evaluated
$Y_{X_{SAC}/S_{lac}}$	0.28	Evaluated
$Y_{X_{LAB}/S_{Rf}}$	0.55	Evaluated
$Y_{X_{SAC}/S_{Rf}}$	0.32	Evaluated
α_{Eth}	1.0	Evaluated
α_{Rf}	0.75	Evaluated
β_{Eth}	0.01	Evaluated
β_{Rf}	0.01	Evaluated
$\mu_{\max LAB_{lac}}$	0.75	Evaluated
$\mu_{\max LAB_{Rf}}$	1.0	Evaluated
$\mu_{\max SAC_{lac}}$	0.48	Evaluated
$\mu_{\max SAC_{Rf}}$	0.55	Evaluated

TABLE 9.2

The Correlation Coefficients and the Confidence Level of Regression Equations Used for Parameter Estimations

Parameters	Correlation Coefficient	Confidence Interval % $100*(1-\alpha)$ (Student t-test)
$\mu_{\max LAB_{lac}}, K_{S_{LAB_{lac}}}, K_{LA_{lac}}$	0.97	97
$\mu_{\max LAB_{Rf}}, K_{S_{LAB_{Rf}}}, K_{LA_{Rf}}$	0.97	97
$\mu_{\max SAC_{lac}}, K_{S_{SAC_{lac}}}$	0.97	97
$\mu_{\max SAC_{Rf}}, K_{S_{SAC_{Rf}}}$	0.97	97
$\alpha_{Eth}, \beta_{Eth}$	0.99	99
α_{Rf}, β_{Rf}	0.99	99

From Table 9.1, it is noted that maximum specific growth rate based on either lactose or riboflavin for *Lactobacillus casei (JUCHE 2)* is higher than *Saccharomyces cerevisae (JUCHE 3)*. It signifies that faster growth rate, and low cell doubling time of *Lactobacillus casei (JUCHE 2)* than *Saccharomyces cerevisae (JUCHE 3)* in presence of lactose. Both consortia can assimilate riboflavin for their growth, but maximum specific growth rate based on riboflavin is lower than lactose for both consortia. This may be justified by the fact that consortia may utilize riboflavin as a secondary nutrient source. Similarly, the biomass yield coefficient based on riboflavin is lower than lactose for two proposed consortia. The value of the kinetic constant of the Luedeking Piret model, β_{Rf} is higher compared to other kinetic constant α_{Rf}, which signifies that the formation of riboflavin is not growth-associated. On the other hand, value of kinetic constant of α_{Eth} is higher compared to β_{Eth}, which implies that the synthesis of ethanol is growth-associated.

Experimental data obtained during the growth phase are being used for comparison with simulated data. In Figure 9.2, the simulated and experimental data have been compared for time histories of concentrations of biomass of *Lactobacillus casei (JUCHE 2)* and *Saccharomyces cerevisae (JUCHE 3)*. From the analysis of the figures, it is evident that the concentrations of biomass are increasing with batch time progress. It is observed that growth of *Lactobacillus casei (JUCHE 2)* is high compared to *Saccharomyces cerevisae (JUCHE 3)*. This may be justified by the fact that low cell doubling time as well as high specific growth rate of *Lactobacillus casei (JUCHE 2)* than other consortium. In both cases, the concentrations of biomass have increased monotonically, and after a certain time it becomes steady with time parameters.

In Figure 9.3, simulated and experimental results of concentration profiles of lactic acid formation have been plotted as a function of reaction time. The experimental and simulated trends are in agreement. Production of lactic acid increases monotonically, and after a certain period it becomes steady with time axis.

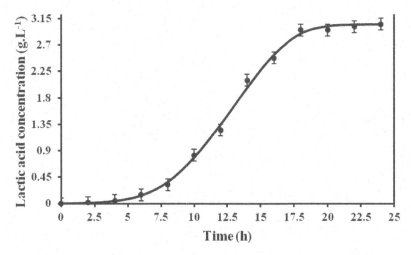

FIGURE 9.2 Simulated (lines) and experimental (points) time histories of concentration of lactic acid in growth medium. Points: Concentration of lactic acid at simulated whey medium (■), concentration of lactic acid at real casein whey medium (•).

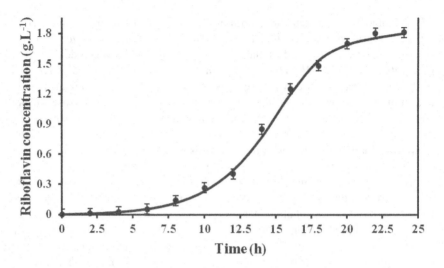

FIGURE 9.3 Simulated (lines) and experimental (points) time histories of concentration of riboflavin in the growth medium. Points: Concentration of riboflavin at simulated whey medium (■), concentration of riboflavin at real casein whey medium (•).

In Figure 9.4, time history of concentration profiles of riboflavin has been plotted as a function of reaction time. In the same figure simulated, and experimental results are superimposed. Proposed model equation is established by good agreement of experimental and the simulated results. It is observed that synthesis of riboflavin increases monotonically and after a certain period it becomes steady with time axis.

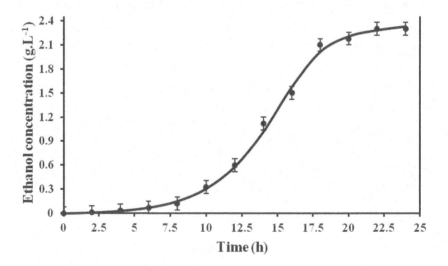

FIGURE 9.4 Simulated (lines) and experimental (points) time histories of concentration of ethanol in the growth medium. Points: Concentration of ethanol at simulated whey medium (■), concentration of ethanol at real casein whey medium (•).

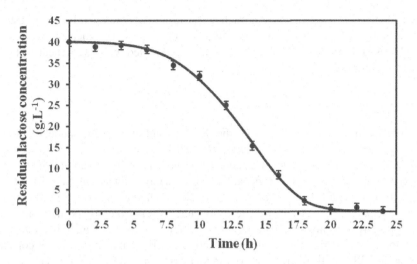

FIGURE 9.5 Simulated (lines) and experimental (points) time histories of concentration of residual lactose in growth medium. Points: concentration of residual lactose at simulated whey medium (■), concentration of residual lactose at real casein whey medium (•).

In Figure 9.5, simulated, and experimental results of concentration profiles of ethanol have been plotted as a function of reaction time. It is observed that with batch time progress, production of ethanol is increased monotonically, and after a certain period it becomes saturated. In this chapter, mostly time histories of simulated and experimental results of concentration profiles of residual lactose have been depicted. Declining patterns of time histories of lactose over the entire period establish the fact that the lactose is assimilated simultaneously by *Lactobacillus casei (JUCHE 2)* and *Saccharomyces cerevisae (JUCHE 3)*. The comparison suggests that simulated trends are able to predict the experimental one. From the analysis of the figures, it is evident that the model can adequately predict the experimental data in all cases. The values of significance level and MS_E for each case have been provided in Table 9.3.

TABLE 9.3

Statistical Evaluation of the Fits of Structured Models to Measurements

Case	MS_E	Confidence Level α (From Fisher F test)	% 100(1 – α)
Biomass concentration of *Lactobacillus casei (JUCHE 2)*	3.25×10^{-2}	0.05	95
Biomass concentration of *Saccharomyces cerevisae (JUCHE 3)*	2.55×10^{-2}	0.048	95.2
Lactic acid concentration	9.65×10^{-7}	0.02	98
Riboflavin concentration	9.88×10^{-7}	0.01	99
Ethanol concentration	9.75×10^{-7}	0.015	98.5
Residual lactose concentration	9.85×10^{-7}	0.01	99

The model may be particularly attractive because it is developed based on established biological principles involving almost no adjustable parameters. The model may be used to predict the behavior of a storage tank fed with sweet meat whey in practical situations.

9.5 CONCLUSION

The present investigation deals with the development of a kinetic model of mixed consortia, namely, *Lactobacillus casei (JUCHE 2)* and *Saccharomyces cerevisae (JUCHE 3)*, grown in sweet meat casein whey. Since this is the first attempt to develop the kinetic model of mixed consortia, a few reasonably valid assumptions have been made during the development of model equations. The model equations contain no adjustable parameters. All the kinetic parameters were evaluated experimentally and through regression analysis, and good correlation indices and confidence intervals were found. Time histories of biomass, lactic acid, riboflavin, and ethanol concentration could be well-predicted by the model equations with very high confidence levels. There is a scope for the modification of proposed model equations by incorporating contamination of other microorganisms, segregation with respect to the age of culture, etc. The model may be used as a unique tool to study the microbe-microbe interaction in natural systems. The present investigation may serve a high impact to the scientific community, owing to disposal of casein whey.

ACKNOWLEDGMENTS

The first author acknowledges the Council of Scientific and Industrial Research (CSIR), New Delhi, India for providing the research fellowship. The reported work is a part of a University Grants Commission (UGC) major project, entitled "Production and Purification of β-galactosidase from Milk Whey-based Lactic Acid Bacteria using Fermentation and Membrane-based Separation Techniques." The contribution of UGC is gratefully acknowledged by all the authors. The authors gratefully acknowledge the academic advice rendered by Prof. (Dr.) Pinaki Bhattacharya, retired professor of Chemical Engineering Department of Jadavpur University, Kolkata, India.

REFERENCES

1. Andersen, J.H., Jenssen, H., Gutterberg, T.J., 2003. Lactoferrin and lactoferricin inhibit Herpes simplex 1 and 2 infection and exhibit synergy when combined with acyclovir. Antiviral Res 58, 209–215.
2. Aschaffenburg, R., Drewry, J., 1957. Improved method for the preparation of crystalline β-lactoglobulin and α-lactalbuminfrom cow's milk. Biochem J 65, 273.
3. Bellamy, W., Takase, M., Yamauchi, K., Wakabayashi, H., Kawase, K., Tomita, M., 1992a. Identification of the bactericidal domain of lactoferrin. Biochim Biophys Acta 1121, 130–136.
4. Berrocal, R., Chanton, S., Juilleart, M.A., Pavillard, B., Scherz, J.C., Jost, R., 1989. Tryptic phosphor peptides from whole casein. II Physicochemical properties related to the solubilization ofcalcium. J Dairy Res 56, 335–341.

5. Boza, J.J., Martinez-Augustin, O., Gil, A., 1995. Nutritional and antigenic characterization of an enzymatic whey protein hydrolysate. J Agric Food Chem 43, 872–875.
6. Cassens, P.W.J.R., Visser, S., Gruppen, H., Voragen, A.G.J., 1999. Lactoglobulin hydrolysis. 1. Peptide composition and functional properties of hydrolysates obtained by the action of plasmin, trypsin and Staphylococcus aureus V8 protease. J Agric Food Chem 47, 2973–2979.
7. Cheison, S.C., Wang, Z., Xu, S.Y., 2007. Preparation of whey protein hydrolysates using a single-and two-stage enzymatic membrane reactor and their immunological and antioxidant properties: Characterization by multivariate data analysis. J Agric Food Chem 55, 3896–3904.
8. Didelot, S., Bordenave-Juchereau, S., Rosenfeld, E., Piot, J.-M., Sannier, F., 2006. Peptides released from acid goat whey by a yeast lactobacillus association isolated from cheese microflora. J Dairy Res 6, 163–170.
9. Ergüder, T.H., Tezel, U., Güven, E., Demirer, G.N., 2001. Anaerobic biotransformation and methane generation potential of cheese whey in batch and UASB reactors. Waste Manage 21, 643–650.
10. Gagnaire, V., Pierre, A., Molle, D., Leonil, J., 1996. Phosphopeptides interact with colloidal calcium phosphate isolated by tryptic hydrolysis of bovine casein micelles. J Dairy Res 63, 405–422.
11. Halken, S., Hansen, K.S., Jacobsen, H.P., Estmann, A., Faelling, A.E., Hansen, L.G., Kier, S.R., Lassen, K., Lintrup, M., Mortensen, M., Ibsen, K.K., Osterballe, O., Host, A., 2000. Comparison of a partially hydrolyzed infant formula with two extensively hydrolyzed formulas for allergy prevention: A prospective randomized study. Pediatr. Allergy Immunol 11, 149–161.
12. Hamer, J., Haheim, H., Gutterberg, T.J., 2000. Bovine lactoferrin is more efficient than bovine lactoferricin in inhibiting HSV-I/II replication in vitro. In: Shimazaki, K. (Ed.), Lactoferrin: Structure, Functions and Applications. Elsevier, Amsterdam, pp. 239–243.
13. Ikeda, M., Nozaki, A., Sugiyama, K., Tanaka, T., Naganuma, A., Tanaka, K., Sekihara, H., Shimotohno, K., Saito, M., Kato, N., 2000. Characterization of antiviral activity of lacto ferrinagainst hepatitis C virus infection in human cultured cells. Virus Res 66, 51–63.
14. Jacquot, A., Gauthier, S.F., Drouin, R., Boutin, Y., 2010. Proliferative effects of synthetic peptides from beta-lactoglobulin and alpha-lactalbumin on murine splenocytes. Int Dairy J 20, 514–521.
15. Juilleart, M.A., Baechler, R., Berrocal, R., Chanton, S., Scherz, J.C., Jost, R., 1989. Tryptic phosphopeptides from whole casein I. Preparation and analysis by fast protein liquid chromatography. J Dairy Res 56, 603–611.
16. Kourkutas, Y., Xolias, V., Kallis, M., Bezirtzoglu, E., Kanellaki, M., 2005. Lactobacillus casei immobilization on fruit pieces for prebiotic additive, fermented milk and lactic acid production. Process Biochem 40, 411–416.
17. Lamas, E.M., Barros, R.M., Balcão, V.M., Malcata, F.X., 2001. Hydrolysis of whey proteins by proteases extracted from Cynara cardunculus and immobilized onto highly activated supports. Enzyme Microbial Technol 28, 642–652.
18. Mahaut, M., Maubois, J.L., Zink, A., Pannetier, R., Veyre, R., 1982. Eléments de fabrication de fromages frais par ultrafiltration sur membrane de coagulum de lait. Tech Lait 961, 9–13.
19. Manso, M.A., López-Fandiño, R., 2003. Angiotensin I converting enzyme-inhibitory activity of bovine, ovine and caprinekappa-casein macro peptides and their tryptic hydrolysates. J Food Prot 66, 1686–1692.
20. Maubois, J.L., 1986. Recent developments of ultrafiltration in dairy industries. In: Drioli, E., Nakagaki, M. (Eds.), Membrane and Membrane Process. Springer, New York, NY, pp. 255–262.

21. McHugh, T.H., Aujurd, J.F., Krochta, J.M., 1994. Plasticized whey protein edible films: Water vapor permeability properties. J Food Sci 59, 416–419.
22. Meisel, H., Schlimme, E., 1996. Bioactive peptides derived from milk proteins: Ingredients for functional foods? Kieler Milchwirts. Forsch 48, 343–357.
23. Modler, H.W., Emmons, D.B., 1977. Properties of whey protein concentrate prepared by heating under acidic conditions. J Dairy Sci. 60 (2), 177–184.
24. Mullally, M.M., Meisel, H., FitzGerald, R.J., 1997. Angiotensin-I-converting enzyme inhibitory activities of gastric and pancreatic proteinase digests of whey proteins. Int Dairy J 7, 299–303.
25. Nagaoka, S., Futumura, Y., Miwa, K., Awano, T., Yamauchi, K., Kanamaru, Y., Tadashi, K., Kuwata, T., 2001. Identification of novel hypocholesterolemic compound derived from bovine milk beta-lactoglobulin. Biochem. Biophys. Res. Commun 281, 11–17.
26. Nath, A., Chakraborty, S., Bhattacharjee, C., Chowdhury, R., 2014. Studies on the separation of proteins and lactose from casein whey by cross-flow ultrafiltration. Desalination and Water Treatment, 54, 481–501.

Index

Printed in the United States
by Baker & Taylor Publisher Services